Mathematical Research

Splines in Numerical Analysis

edited by
J.W. Schmidt
H. Späth

Volume 52

AKADEMIE-VERLAG BERLIN

In this series original contributions of mathematical research in all fields are contained, such as

— research monographs
— collections of papers to a single topic
— reports on congresses of exceptional interest for mathematical research.

This series is aimed at promoting quick information and communication between mathematicians of the various special branches.

In diese Reihe werden Originalbeiträge zu allen Gebieten der mathematischen Forschung aufgenommen wie

— Forschungsmonographien
— Sammlungen von Arbeiten zu einem speziellen Thema
— Berichte von Tagungen, die für die mathematische Forschung besonders aktuell sind.

Die Reihe soll die schnelle Information und gute Kommunikation zwischen den Mathematikern der verschiedenen Fachgebiete fördern.

Manuscripts
in English and German comprising at least 100 pages and not more than 500 pages can be admitted to this series. With respect to a quick publication the manuscripts are reproduced photomechanically. Authors who are interested in this series please turn directly to the 'Akademie-Verlag'. Here you will get more detailed information about the form of the manuscripts and the modalities of publication.

Manuskripte
in englischer und deutscher Sprache, die mindestens 100 Seiten und nicht mehr als 500 Seiten umfassen, können in diese Reihe aufgenommen werden. Im Interesse einer schnellen Publikation werden die Manuskripte auf fotomechanischem Weg reproduziert. Autoren, die an der Veröffentlichung entsprechender Arbeiten in dieser Reihe interessiert sind, wenden sich bitte direkt an den Akademie-Verlag. Sie erhalten dort genauere Informationen über die Gestaltung der Manuskripte und die Modalitäten der Veröffentlichung.

Splines in Numerical Analysis

Mathematical Research · Mathematische Forschung

Wissenschaftliche Beiträge
herausgegeben von der
Akademie der Wissenschaften der DDR
Karl-Weierstraß-Institut für Mathematik

Band 52

Splines in Numerical Analysis

Splines in Numerical Analysis

Contributions to the International Seminar ISAM–89
held in Weissig (GDR), April 24–28, 1989

edited by

Jochen W. Schmidt
Helmuth Späth

Akademie-Verlag Berlin 1989

Herausgeber:

Prof. Dr. Jochen W. Schmidt
Technische Universität Dresden
Sektion Mathematik

Prof. Dr. Helmuth Späth
Universität Oldenburg
Fachbereich Mathematik

Die Titel dieser Schriftenreihe werden vom Originalmanuskript der Autoren reproduziert.

ISBN 3-05-500679-8
ISSN 0138-3019

Erschienen im Akademie-Verlag Berlin, Leipziger Str. 3-4, Berlin, DDR-1086

Lizenznummer: 202·100/412/89
Printed in the German Democratic Republic
Gesamtherstellung: VEB Kongreß- und Werbedruck, Oberlungwitz, DDR-9273
Lektor: Dr. Reinhard Höppner
LSV 1065
Bestellnummer: 763 990 1 (2182/52)

02800

PREFACE

This volume compiled in preparation of the workshop ISAM 89 collects survey and more extensive original articles on up-to-date problems in the application of splines in Numerical Analysis. It is intended to let appear the volume already at the beginning of the seminar, in April 1989. All the published articles have been subjected to a reviewing process.

The papers of this volume are assumed to be of interest for a large number of scientists in academic and industrial research. In the centre there are contributions to the following topics: Data smoothing by splines, shape preserving spline interpolation, surface approximation by means of splines as well as applications of splines to the solution of differential and integral equations covering singularly perturbed boundary value problems, singular integral equations, collocation methods and enclosing techniques.

The editors are indebted to all the authors who contributed to these proceedings, and to the referees for their careful cooperation. They sincerely thank Dr.R.Höppner of the Akademie-Verlag, Berlin for his kind expressions of interest and encouragement.

J.W.Schmidt
Dresden

H.Späth
Oldenburg

January 1989

CONTENTS

A DIRECT CUBIC SPLINE APPROACH FOR INITIAL VALUE PROBLEMS

Mohamed Naim Yehia Anwar [1]

1. Introduction

Numerical methods for ordinary differential equations are in general divided into discrete andcontinuous ones. Among such continuous methods are those that use splines. Such methods are considered as well established approach for solving ordinary differential equations (see for example [3],[4], and [5]).

In the present work a direct cubic spline approach for approximating the solution of initial value problems governed by ordinary differential equations is formulated. The general outline of the resulting algorithm is similar to that given in [3]; nevertheless the construction of the spline approximation is different and furthermore error bounds for the solution and its first three derivatives are derived.

2. The Considered Cubic Spline

Let $\{x_i, i=0,1,..,N\}$ be a uniform partition of the interval $[a,b]$. Denote by $S_{N,3}^{(2)}$ the linear space of cubic splines s such that $s \in C^2[a,b]$. s is a cubic polynomial in each subinterval $[x_i, x_{i+1}]$. Set $h = x_{i+1}-x_i$ $(i=0,1,..,N-1)$. If Φ is a real valued function defined in $[a,b]$ then Φ_i stands for $\Phi(x_i)$, $(i=0,1,..,N)$.

Proposition 1: There exists a unique cubic spline $s \in S_{N,3}^{(2)}$ whose first derivatives s_i' $(i=0,1,..,N)$ are given along with s_o and s_o''. This spline is defined in $[x_i, x_{i+1}]$ by

$$s(x) = s_i W_o(t) + h s_i' W_1(t) + h^2 s_i'' W_2(t) + h s_{i+1}' W_3(t) \tag{2.1}$$

where

$$W_o(t) = 1,\ W_1(t) = t - \frac{t^3}{3},\ W_2(t) = \frac{t^2}{2} - \frac{t^3}{3},\ W_3(t) = \frac{t^3}{3} \tag{2.2}$$

and $t = (x-x_i)/h$.

The coefficients s_i, s_i'' in (2.1) are given by the recurrence formulae $(i=1,2,..,N)$

$$s_i'' = -s_{i-1}'' + \frac{2}{h}(-s_{i-1}' + s_i')$$

$$s_i = s_{i-1} + \frac{h}{3}(2s_{i-1}' + s_i') + \frac{h^2}{6} s_{i-1}'' \tag{2.3}$$

s_o, s_o'' are known.

---

[1] Department of Engineering Mathematics and Physics, Faculty of Engineering, Alexandria University, Alexandria, EGYPT

Proof: If $P_3(t)$ is a cubic polynomial in [0,1] then it can be expressed as

$$P_3(t) = P_3(0)W_0(t)+P_3'(0)W_1(t)+P_3''(0)W_2(t)+P_3'(1)W_3(t).$$

Using $P_3(t) = 1$, t, t^2 and t^3, the corresponding expressions in (2.2) for W_0, W_1, W_2 and W_3 can be obtained from the resulting linear system of equations.

Now for a fixed $i \in \{0,1,..,N-1\}$, set $x = x_i+th$, $0 \leq t \leq 1$ and thus we get the expression of s given in (2.1). We have an analogous expression for s in $[x_{i-1},x_i]$. Since $s \in C^2[a,b]$, so the continuity conditions $s(x_i-0) = s(x_i+0)$ and $s''(x_i-0) = s''(x_i+0)$ lead to the above recurrence formulae in (2.3). This completes the proof.

3. An algorithm for Initial Value Problems

Consider the initial value problem

$$y' = g(x,y), \quad a \leq x \leq b, \quad y(a) \text{ is given.} \tag{3.1}$$

We suppose that g is defined and continuous in $[a,b] \times R$ and it has a continuous bounded derivative with respect to y. That is

$$|g_y(x,y)| \leq L \text{ in } [a,b] \times R. \tag{3.2}$$

This guarantees the existence and uniqueness of the solution denoted by $y = f(x)$.

We want to approximate $y = f(x)$ by the cubic spline (2.1)-(2.3) of section 2.

The solution $y = f(x)$ of the initial value problem (3.1) is approximated in $[a,b]$ by the cubic spline $s \in S^{(2)}_{N,3}$ defined by the following algorithm:

1) Compute s_0', s_0'' from

$$s_0' = y'(a) = g(a,s_0) \tag{3.3.1}$$

$$s'' = y''(a) = g_x(a,s_0) + g_y(a,s_0)s_0' \tag{3.3.2}$$

where $s_0 = y(a)$.

2) For i=1,2,..,N compute

$$s_i' = g(x_i,s_i)$$

$$s_i = s_{i-1} + \frac{h}{3}(2s_{i-1}'+s_i') + \frac{h^2}{6}s_{i-1}'' \tag{3.4}$$

$$s_i'' = -s_{i-1}'' + \frac{2}{h}(-s_{i-1}'+s_i').$$

This is equivalent to

$$s_i = s_{i-1} + \frac{h}{3}[2g(x_{i-1},s_{i-1})+g(x_i,s_i)] + \frac{h^2}{6}s_{i-1}'' \tag{3.5.1}$$

$$s_i'' = -s_{i-1}'' + \frac{2}{h}[-g(x_{i-1},s_{i-1}) + g(x_i,s_i)]. \tag{3.5.2}$$

3) In each subinterval $[x_i, x_{i+1}]$, i=0,1,..,N-1, let

$$s(x) = s_i W_0(t) + h s_i' W_1(t) + h^2 s_i'' W_2(t) + h s_{i+1}' W_3(t) \quad (3.6)$$

where $t = (x-x_i)/h$, W_0, W_1, W_2, W_3 are given by (2.2).

Notice that if g is nonlinear with respect to its argument y then (3.5.1) is nonlinear in s_i. The solution is obtained by the fixed point iteration

$$s_i^{(p+1)} = s_{i-1} + \frac{h}{3}\,[2g(x_{i-1}, s_{i-1}) + g(x_i, s_i^{(p)})] + \frac{h^2}{6}\, s_{i-1}''. \quad (3.7)$$

It will be shown (see section 4) that this iteration converges for any $s_i^{(0)}$ to the same limit provided that $h < 3/L$.

4. Convergence of the Algorithm and Error Bounds

In this section we give L_∞-bounds for s(x)-f(x) and its first three derivatives in [a,b]. The following analysis is carried out for the general nonlinear case. $\|\cdot\|$ denotes the L_∞-norm, $e^{(r)}(x) = s^{(r)}(x) - f^{(r)}(x)$, r=0,1,2,3, where r denotes the order of differentiation.

Lemma 1: Let s_i and s_i' (i=0,1,..,N) be given by (3.4) then

$$|e_i'| \leq L|e_i|. \quad (4.1)$$

Proof: We have from (3.1) and (3.4)

$$s_i' = g(x_i, s_i), \quad f_i' = g(x_i, f_i).$$

Thus by the mean value theorem

$$s_i' - f_i' = g_y(x_i, \delta_i)(s_i - f_i)$$

where δ_i is between s_i and f_i. Hence (4.1) follows from the condition (3.2).

Lemma 2: If $h < 3/L$ then

$$|e_1| \leq \frac{9h^4}{64(3-hL)}\, \|f^{(4)}\| \quad (4.2)$$

provided $f \in C^4[a,b]$.

Proof: Equation (3.4) can be rearranged in the following form

$$s_1 = [s_0 + \frac{h}{3}(2f_0' + f_1') + \frac{h^2}{6} s_0''] + \frac{h}{3}(2e_0' + e_1').$$

Since $s_0 = f_0$ and $s_0'' = f_0''$ from (3.3), thus by subtracting f_1 from both sides of the above equality and expanding the right hand side about $x_0 + h/2$ using Taylor's expansion of order 4 we get

$$|e_1| \le \frac{3h^4}{64}||f^{(4)}|| + \frac{h}{3}(2|e_0'|+|e_1'|).$$

Using (4.1) and $e_0' = 0$ we obtain (4.2).

Theorem 1: Let s_i $(i=0,1,..,N)$ be given by (3.4). For $h < 3/L$ and $f \in C^5[a,b]$ we have the following error bound $(i=1,2,..,N)$

$$|e_i| \le \frac{9h^4}{64(3-hL)} \|f^{(4)}\|\hat{E}_i + \frac{h^4}{180L}\|f^{(5)}\| (\hat{E}_i-1),$$

where $\hat{E}_i = \exp[\frac{6L(x_i-a)}{3-hL}]$. (4.3.1)

Furthermore if $h < 2/L$ then (4.3.1) leads to

$$|e_i| \le (\frac{9}{64}\|f^{(4)}\| + \frac{1}{180L}\|f^{(5)}\|)h^4 \exp[6L(b-a)] - \frac{h^4}{180L}\|f^{(5)}\|. \quad (4.3.2)$$

Proof: From (3.4) we have $(i=1,2,..,N-1)$

$$s_i = s_{i-1} + \frac{h}{3}(2s_{i-1}'+s_i') + \frac{h^2}{6}s_{i-1}''$$

$$s_{i+1} = s_i + \frac{h}{3}(2s_i'+s_{i+1}') + \frac{h^2}{6}s_i''.$$

Adding these two equations and using the third equation of (3.4) we get $(i=1,2,..,N-1)$

$$s_{i+1} = s_{i-1} + \frac{h}{3}(s_{i-1}'+4s_i'+s_{i+1}') \quad (4.4)$$

which can be rewritten as

$$s_{i+1} = e_{i-1}+f_{i-1} + \frac{h}{3}(f_{i-1}'+4f_i'+f_{i+1}') + \frac{h}{3}(e_{i-1}'+4e_i'+e_{i+1}').$$

Subtracting f_{i+1} from both sides of the above equation and noticing that

$$f_{i-1}-f_{i+1} + \frac{h}{3}(f_{i-1}'+4f_i'+f_{i+1}') = -\int_{x_{i-1}}^{x_{i+1}} f'(u)du + \frac{h}{3}(f_{i-1}'+4f_i'+f_{i+1}')$$

which is the error term of the classical Simpson's rule over two sub-intervals, we obtain

$$|e_{i+1}| \le |e_{i-1}| + \frac{h^5}{90}\|f^{(5)}\| + \frac{h}{3}(|e_{i-1}'|+4|e_i'|+|e_{i+1}'|). \quad (4.5)$$

Using (4.1) and since $h < 3/L$ we get

$$|e_{i+1}| \le \alpha|e_i| + \beta|e_{i-1}| + \sigma \quad (4.6)$$

where

$$\alpha = \frac{4hL}{(3-hL)}, \quad \beta = \frac{(3+hL)}{(3-hL)}, \quad \text{and } \sigma = \frac{h^5 f^{(5)}}{[30(3-hL)]}. \quad (4.7)$$

Then (4.6) is equivalent to

$$E_{i+1} \le A\, E_i + \sigma V \tag{4.8}$$

where

$$E_i = \begin{bmatrix} |e_i| \\ |e_{i-1}| \end{bmatrix}, \quad i=1,2,\ldots,N, \quad A = \begin{bmatrix} \alpha & \beta \\ 1 & 0 \end{bmatrix}, \quad V = \begin{bmatrix} 1 \\ 0 \end{bmatrix}.$$

From (4.8) we can arrive at

$$\|E_{i+1}\| \le \|A\|^i \|E_1\| + \sigma(\|A\|^i - 1)/(\|A\| - 1). \tag{4.9}$$

But

$$\|A\| = \alpha + \beta = (3+5hL)/(3-hL) = 1 + 6hL/(3-hL).$$

Thus

$$\|A\|^i = [1+6hL/(3-hL)]^i \le \exp[6ihL/(3-hL)]. \tag{4.10}$$

But $\|E_1\| = \max(|e_0|, |e_1|) = |e_1|$ since $e_0 = 0$. Hence (4.9) and (4.10) give

$$\|e_i\| \le |e_1| \hat{E}_i + \frac{\sigma(3-hL)}{6hL} (\hat{E}_i - 1).$$

Using (4.2) and (4.7) we arrive at the error bound (4.3.1). When $hL < 2$ that is $3-hL > 1$ and since $x_i - a \le b-a$ then (4.3.2) follows. This completes the proof.

Notice that an error bound on e' can be obtained by multiplying that for e by L (Lemma 1).

Lemma 3: If $h < 3/L$ then

$$|e_1''| \le \left[\frac{1}{6} + \frac{9hL}{32(3-hL)}\right] h^2 \|f^{(4)}\| \tag{4.11}$$

provided $f \in C^4[a,b]$.

Proof: We have $s_0' = f_0'$, $s_0'' = f_0''$. For $i=1$, (3.4) gives

$$s_1'' - f_1'' = -f_0'' - f_1'' + \frac{2}{h}(-f_0' + f_1') + \frac{2}{h}(-e_0' + e_1')$$

or

$$e_1'' = -f_0'' - f_1'' + \frac{2}{h} \int_a^{a+h} f''(t)dt + \frac{2}{h}(-e_0' + e_1') .$$

Integrating by parts twice we get

$$e_1'' = \int_a^{a+h} \left[-\frac{h}{4} + \frac{1}{h}\left(t - \frac{h}{2}\right)^2\right] f^{(4)}(t)dt + \frac{2}{h}(-e_0' + e_1') .$$

Now by the mean value theorem for integrals (4.1) and (4.2) we obtain

$$|e_1''| \le \frac{h^2}{6}\|f^{(4)}\| + \frac{9h^3L}{32(3-hL)} \|f^{(4)}\|$$

which is (4.11).

Theorem 2: Let s''_i $(i=0,1,\dots,N)$ be given by (3.4). For $h<\frac{3}{L}$ and $f \in C^5[a,b]$ we have the following error bound $(i=1,2,\dots,N)$

$$|e''_i| \le |e''_1| + (x_i-a)\left[\frac{17}{12}h^2\|f^{(5)}\| + \frac{4L}{h^2}\|e\|\right]$$

where $|e''_1|$ satisfies (4.11), $\|e\| = \max_{1\le i\le N}|e_i|$ and $|e_i|$ satisfies (4.3.1).

Proof: By (3.4) we can write $(i=1,2,\dots,N-1)$

$$s''_i = \frac{2}{h}(-s'_{i-1} + s'_i) - s''_{i-1}$$

and $$s''_{i+1} = \frac{2}{h}(-s'_i + s'_{i+1}) - s''_i \;.$$

Substituting the first equation into the second one and rearranging, we arrive at

$$e''_{i+1} = e''_{i-1} + f''_{i-1} - f''_{i+1} + \frac{2}{h}(f'_{i-1} - 2f'_i + f'_{i+1}) + \frac{2}{h}(e'_{i-1} - 2e'_i + e'_{i+1}) \;.$$

Expanding about x_{i-1} using Taylor's expansion of order 5, we can show that $(i=1,2,\dots,N-1)$

$$|e''_{i+1}| \le |e''_{i-1}| + \frac{17}{6}h^3\|f^{(5)}\| + \frac{8L}{h}\|e\| \;. \tag{4.13}$$

Now, (4.12) is obtained as solution of this linear difference inequality.

From (3.6) we have for x restricted to $[x_i, x_{i+1}]$ $(i=0,1,\dots,N-1)$

$$s'''(x) = -\frac{2}{h^2}s'_i - \frac{2}{h}s''_i + \frac{2}{h^2}s'_{i+1} \tag{4.14}$$

which is constant over this subinterval.

Lemma 4: For $i=0,1,2,\dots,N$ we have

$$|e'''_i| \le \frac{h}{3}\|f^{(4)}\| + \frac{4L}{h^2}\|e\| + \frac{2}{h}\|e''\| \tag{4.15}$$

provided $f \in C^5[a,b]$ and s'''_i is defined as in (4.14).

Proof: From (4.14) we can write

$$e_i''' = -f_i''' - \frac{2}{h^2} f_i' - \frac{2}{h} f_i'' + \frac{2}{h^2} f_{i+1}' - \frac{2}{h^2} e_i' - \frac{2}{h} e_i'' + \frac{2}{h^2} e_{i+1}'.$$

Expanding the right hand side using Taylor's expansion of order 5 about x_i we get the result (4.15).

Theorem 3: Let s(x) be the cubic spline defined in (3.6). If $h < 3/L$ and $f \in C^5[a,b]$ then for any $x \in [a,b]$ we have

$$|e^{(r)}(x)| \le \begin{cases} (1+hL)\|e\| + \frac{h^2}{6}\|e''\| + \frac{7h^4}{72}\|f^{(4)}\| & r=0 \\ 5L\|e\| + h\|e''\| + \frac{5}{6}h^3\|f^{(4)}\| & r=1 \\ \frac{4L}{h}\|e\| + \|e''\| + \frac{5}{6}h^2\|f^{(4)}\| & r=2 \\ \frac{4L}{h^2}\|e\| + \frac{2}{h}\|e''\| + \frac{4h}{3}\|f^{(4)}\| & r=3 \end{cases} \quad (4.16)$$

where $e^{(r)}(x) = s^{(r)}(x) - f^{(r)}(x)$, $\|e\|$ and $\|e''\|$ are bounded by (4.3.1) and (4.12), respectively.

Proof: Subtracting f(x) from both sides of (3.6) and expanding the right hand side about x_i using Taylor's expansion of order 4 we get

$$e(x) = e_i + h^2\left(\frac{t^2}{2} - \frac{t^3}{3}\right)e_i'' + h\left[\left(t - \frac{t^3}{3}\right)e_i' + \frac{t^3}{3}e_{i+1}'\right] +$$

$$+ h^4\left[\frac{t^3}{18} f^{(4)}(\mu_i) - \frac{t^4}{24} f^{(4)}(\delta_i)\right].$$

Using (4.1) and since $0 \le t \le 1$ we can establish the first result in (4.16).

From (3.6) we obtain

$$s''(x) = \frac{s_i'}{h}(-2t) + s_i''(1-2t) + \frac{s_{i+1}'}{h}(2t).$$

Subtracting $f''(x)$ from both sides and expanding about x_i as above one can arrive at the third result in (4.16) (r=2).

Similarly the fourth result that corresponds to r=3, follows from (4.14) by subtracting $f'''(x)$ from both sides and expanding as above.

Finally we can write

$$s'(x) - f'(x) = \int_{x_i}^{x} [s''(\theta) - f''(\theta)]d\theta + (s_i' - f_i')$$

from which we have

$$|e'(x)| \le \int_{x_i}^{x} |e''(\theta)|d\theta + L\|e\|.$$

Using (4.16), r=2, we can prove the second result in (4.16) (r=1). This completes the proof.

5. Convergence of the Associated Fixed Point Iteration

Consider the fixed point iterative process (3.7) which, on writing $v_p = s_i^{(p)}$ (i is fixed) may be expressed as

$$v_{p+1} = \Phi(v_p) \tag{5.1}$$

where Φ maps R into R and Φ is contractive for $h < 3/L$ since from (3.2) and (3.7) it can be shown that

$$|\Phi(u)-\Phi(v)| = \frac{h}{3} |g(x_i,u)-g(x_i,v)| \leq \frac{hL}{3} |u-v|.$$

Note that $h < 3/L$ is the same condition as was necessary for the validity of all the previous error bounds. It follows then from the classical fixed point problem and associated iteration that (5.1) converges to its unique fixed point (which is s_i) for any starting value $s_i^{(o)}$. A suitable choice would be $s_i^o = s_{i-1}$.

6. Numerical Results

In order to demonstrate the performance of the proposed method the following numerical examples are considered:

Example 1

$$y' = \frac{1}{x^2} - \frac{y}{x} - y^2, \quad 1 \leq x \leq 2 ,$$

and $y(1) = -1$. The exact solution is $y = f(x) = -1/x$.

Table 1 gives the maximum error bounds for $|s_i^{(r)}-f^{(r)}(x_i)|$, r=0,1,2 and 3 for different choices of the step size h. For each value of r the entries in the top row are calculated values while those in the second one are the estimated ones based on the theoretical error bounds.

Table 1

Maximum error bounds for $\|s^{(r)}-f^{(r)}\|$	h=0.1	h=0.05	h=0.025	h=0.01
r=0	2.9×10^{-5} 1.2×10^{-1}	2.4×10^{-6} 4.9×10^{-3}	2.4×10^{-7} 2.5×10^{-4}	3.0×10^{-7} 5.8×10^{-6}
r=1	2.6×10^{-5} 7.5×10^{-3}	2.0×10^{-6} 9.8×10^{-4}	1.8×10^{-7} 1.2×10^{-4}	2.4×10^{-7} 1.2×10^{-5}
r=2	3.1×10^{-2} 1.8×10^{0}	8.8×10^{-3} 4.5×10^{-1}	2.3×10^{-3} 1.1×10^{-1}	4.0×10^{-4} 5.3×10^{-2}
r=3	1.2×10^{0} 3.6×10	6.9×10^{-1} 1.8×10	3.7×10^{-1} 9.0×10^{0}	1.6×10^{-1} 3.7×10^{0}

Example 2

$$y' = -x\,y^2\ ,\ 2 \le x \le 3 \quad \text{and } y(2) = 1.$$

The exact solution is $y = f(x) = 2/(x^2-2)$.

Table 2 gives the maximum error bounds for $|s_i^{(r)}-f^{(r)}(x_i)|$, r=0,1,2 and 3 as before.

Table 2

Maximum error bounds for $\|s^{(r)}-f^{(r)}\|$	h=0.1	h=0.05	h=0.025	h=0.01
r=0	3.6×10^{-4} 5.3×10^{3}	2.4×10^{-5} 5.8×10	5.2×10^{-6} 2.0×10^{0}	7.2×10^{-7} 3.0×10^{-2}
r=1	7.8×10^{-4} 8.3×10^{-2}	6.3×10^{-5} 1.0×10^{-2}	1.6×10^{-5} 1.3×10^{-3}	2.0×10^{-6} 9.4×10^{-5}
r=2	4.2×10^{-1} 8.7×10	1.1×10^{-1} 2.2×10	2.8×10^{-2} 6.0×10^{0}	4.3×10^{-3} 1.0×10^{0}
r=3	1.7×10 1.7×10^{3}	1.5×10 8.7×10^{2}	1.4×10 4.3×10^{2}	1.2×10 1.7×10^{2}

The theoretical error bounds given in the previous sections are rough estimations only establishing the order of convergence. Improved error inclusions would be attainable by a refined estimation of the Lipschitz constant and of the remainder terms in the Taylor expansions.

The above results show that the calculated error bounds are within the theoretical bounds obtained earlier. Also several other numerical tests have been performed and results are found to be in agreement with the theoretical analysis presented in the previous sections.

References

[1] Anwar, M.N., El Tarazi, M.N.: Direct cubic spline with application to quadrature. Submitted for publication.

[2] Henrici, P.: Discrete variable methods in ordinary differential equations. John Wiley and Sons, Inc. 1962.

[3] Lascalzo, F.R., Talbot, T.D.: Spline function approximations for solutions of ordinary differential equations. SIAM J.Numer.Anal., Vol. 4., No.3, 1967.

[4] Werner, H.: Interpolation and integration of initial value problems of ordinary differential equations by regular splines. SIAM J.Numer.Anal., Vol.12, No.2, 1975.

[5] Werner, H.: Spline functions and the numerical solutions of differential equations. Special Topics of applied mathematics. Proceedings of the seminar held at the GMD, Bonn, 8-1o October 1979. Eds. Frehse, J., Pallaschke, D. and U.Trottenberg. North Holland NY. 1980.

ITERATIVE SPLINE BOUNDS FOR SYSTEMS OF BOUNDARY VALUE PROBLEMS

S. C a r l [1] and C. G r o s s m a n n [2]

1. Introduction

Monotone iteration schemes form a powerful technique as well for the analysis [2], [6], [7] as for the generation of enclosures of the solution [2], [3] of boundary value problems. In the present paper we deal with systems of weakly coupled two-point boundary value problems without monotonicity assumptions w.r.t. the nonlinear right hand side.

Basing on the approach given in [3], [4] here we investigate some new iterative mapping which enables a decoupling of the systems generated by the method. Furthermore we combine this algorithm with monotone discretization as considered in [1],[5]. This leads to iterations in spaces of exponential splines with a fixed finite dimension. Thus the proposed method can be implemented on computers.

We derive some analytical properties and we prove the method to converge. Finally, numerical results are given.

2. Decoupled monotone iterations

Throughout this paper we deal with systems of two-point boundary value problems of the following type:
Find sufficiently smooth functions $u_i: \bar{\Omega} \to R$, $i=1(1)m$ such that

$$\begin{aligned} -u_i'' &= f_i(\cdot,u) \text{ in } \Omega = (a,b) \\ u_i(a) &= u_i(b) = 0 \ , \quad i=1(1)m \end{aligned} \tag{2.1}$$

holds.
Here $\Omega = (a,b)$ and $f_i: \bar{\Omega} \times R^m \to R$ denote some prescribed interval and given continuous functions, respectively.
Next, we sketch the construction of the iteration map used in the technique to be proposed. Let be given some functions

$$\underline{u}, \bar{u} \in C(\bar{\Omega})^m := C(\bar{\Omega}) \times \ldots \times C(\bar{\Omega})$$

satisfying

$$\underline{u}|_\Gamma \leq 0 \leq \bar{u}|_\Gamma \text{ and } \underline{u} \leq \bar{u} \text{ in } \Omega \ , \text{ where } \Gamma := \partial\Omega. \tag{2.2}$$

[1] Carl Schorlemmer Institute of Technology, Department of Mathematics, Otto-Nuschke-Straße, DDR-4200 Merseburg

[2] Dresden University of Technology, Department of Mathematics Mommsenstraße 13, DDR-8027 Dresden

Here and in the sequel we use the natural partial ordering. Now, we relate functions $\underline{v},\bar{v}: \bar{\Omega} \to R^m$ to the interval

$U = [\underline{u},\bar{u}]$ by

$$-\bar{v}_i''(x)+d_i\bar{v}_i(x) = \max_{\underline{u}(x)\leq s\leq \bar{u}(x)} \{d_i s_i+f_i(x,s)\}, \quad x \in \Omega$$

$$\bar{v}_i(a) = \bar{v}_i(b) = 0$$

and (2.3)

$$-\underline{v}_i''(x)+d_i\underline{v}_i(x) = \min_{\underline{u}(x)\leq s\leq \bar{u}(x)} \{d_i s_i+f_i(x,s)\}, \quad x \in \Omega$$

$$\underline{v}_i(a) = \underline{v}_i(b) = 0 .$$

Where $d_i \geq 0$, $i=1(1)m$ denote some reals specified later. Because of the inverse monotonicity of the differential operator

$$[(L+D_i)w](x) := -w''(x)+d_i w(x), \quad x \in \Omega \tag{2.4}$$

with homogenuous first order boundary conditions occuring in (2.3) and

$$\max_{\underline{u}(x)\leq s\leq \bar{u}(x)} \{d_i s_i+f_i(x,s)\} \geq \min_{\underline{u}(x)\leq s\leq \bar{u}(x)} \{d_i s_i+f_i(x,s)\}$$

we have $\underline{v} \leq \bar{v}$. Thus, an operator T mapping intervals $[\underline{u},\bar{u}]$ into intervals $[\underline{v},\bar{v}]$ is defined via (2.3) according to

$$T[\underline{u},\bar{u}] := [\underline{v},\bar{v}]. \tag{2.5}$$

Obviously, the operator T is monotone in the following sense:

$$[\underline{w},\bar{w}] \subset [\underline{u},\bar{u}] \text{ implies } T[\underline{w},\bar{w}] \subset T[\underline{u},\bar{u}]. \tag{2.6}$$

Thus, we can define a monotone iteration scheme by

Algorithm 1:
Step 1: Select some interval $U^o = [\underline{u}^o,\bar{u}^o]$ satisfying (2.2) and

$$U^o \supset T\, U^o. \tag{2.7}$$

Step 2: Determine recursively

$$U^{k+1} := T\, U^k \tag{2.8}$$

for $k=0,1,\dots$.

In contrast to [4] the method given above rests on decoupled systems (2.3) of boundary value problems.

Next, we investigate the convergence properties of method (2.8) with an operator T defined by (2.3),(2.5). Let the right hand sides f_i of system (2.1) be locally Lipschitz continuous functions, i.e.

$$|f_i(x,s)-f_i(y,t)| \leq \ell_{io}|x-y| + \sum_{j=1}^{m} \ell_{ij}|s_j-t_j|$$
(2.9)
$$\text{for any } x,y \in \bar{\Omega};\ s,t \in [\min_{x\in\bar{\Omega}} \underline{u}^o(x), \max_{x\in\bar{\Omega}} \bar{u}^o(x)]$$

holds with constants $\ell_{ij} \geq 0$, $i=1(1)m$, $j=0(1)m$.

Additionally, we denote by ℓ_i, $i=1(1)m$ one sided Lipschitz constants of the functions f_i w.r.t. the i-th argument, i.e.

$$(f_i(x,s)-f_i(x,t))(t_i-s_i) \geq \ell_i(t_i-s_i)^2$$
$$\text{for any } x \in \bar{\Omega};\ s,t \in [\underline{u}^o(x),\bar{u}^o(x)] \text{ with } s_j = t_j,\ j \neq i. \tag{2.10}$$

Obviously, one can select $\ell_i \geq -\ell_{ii}$, $i=1(1)m$.

Using the constants defined by (2.9),(2.10) and the least eigenvalue λ of

$$-\varphi'' = \lambda\varphi \text{ in } \Omega$$
$$\varphi(a) = \varphi(b) = 0$$

an (m,m)-Matrix $A = (a_{ij})$ is defined by

$$a_{ij} = \delta_{ij}(\lambda+\ell_i)-2(1-\delta_{ij})\ell_{ij},\ i,j=1(1)m. \tag{2.11}$$

In the sequel we assume A to be positive definite. We introduce operators $\bar{S}_i, \underline{S}_i$ by

$$[\bar{S}_iU](x) := \max_{s\in U(x)} \{d_is_i+f_i(x,s)\}$$
$$[\underline{S}_iU](x) := \min_{s\in U(x)} \{d_is_i+f_i(x,s)\} \qquad x \in \bar{\Omega}. \tag{2.12}$$

with vector valued interval functions $U = [\underline{u},\bar{u}]$. Using these operators the solutions $\underline{v},\bar{v}$ of system (2.3) are given by

$$\bar{v}_i = (L+D_i)^{-1}\ \bar{S}_iU,$$
$$\underline{v}_i = (L+D_i)^{-1}\ \underline{S}_iU. \qquad i=1(1)m \tag{2.13}$$

Here the boundary conditions are supposed to be included in the definition of $(L+D_i)^{-1}$ due to (2.4). In [1] a strict description of this differential operators using Sobolev spaces is given.

<u>Lemma 1</u>: The mappings $\bar{S}_i,\underline{S}_i : C(\bar{\Omega})^m \times C(\bar{\Omega})^m \to C(\bar{\Omega})$ according to (2.12) are continuous for any $\underline{u},\bar{u} \in U^o$.

<u>Proof</u>: Firstly we show $\bar{S}_iU, \underline{S}_iU \in C(\bar{\Omega})$. With the Lipschitz continuity (2.9) and definition (2.12) we obtain

$$|[\bar{S}_iU](x)-[\bar{S}_iU](y)| \leq \ell_{io}|x-y|+ \sum_{j=1}^{m}(\delta_{ij}d_i+\ell_{ij})(|\bar{u}_j(x)-\bar{u}_j(y)| +$$
$$+|\underline{u}_j(x)-\underline{u}_j(y)|) \text{ for any } x,y \in \bar{\Omega}.$$

Thus $\bar{S}_i U$ forms a continuous function. Secondly we prove $\bar{S}_i U$ to depend continuously on the bounds $\underline{u},\bar{u}$ of the interval U. Let $x \in \bar{\Omega}$ be fixed. By definition (2.12) some $p \in U(x)$ exists such that

$$[\bar{S}_i U](x) = d_i p_i + f_i(x,p).$$

On the other hand we can find some $q \in V(x)$ satisfying

$$|p_i - q_i| \le |\bar{u}_i(x) - \bar{v}_i(x)| + |\underline{u}_i(x) - \underline{v}_i(x)|.$$

With (2.9),(2.12) this leads to

$$[\bar{S}_i U](x) = d_i p_i + f_i(x,p) \le d_i q_i + f_i(x,q) + d_i |p_i - q_i| +$$

$$+ \sum_{j=1}^{m} \ell_{ij} |p_j - q_j| \le [\bar{S}_i V](x) + \sum_{j=1}^{m} (\delta_{ij} d_i + \ell_{ij})(\|\bar{u}_j - \bar{v}_j\|_{C(\bar{\Omega})} + \tag{2.14}$$

$$+ \|\underline{u}_j - \underline{v}_j\|_{C(\bar{\Omega})}.$$

The role of U and V can be interchanged, thus, (2.14) results in

$$\|\bar{S}_i U - \bar{S}_i V\|_{C(\bar{\Omega})} \le \sum_{j=1}^{m} (\delta_{ij} d_i + \ell_{ij})(\|\bar{u}_j - \bar{v}_j\|_{C(\bar{\Omega})} + \|\underline{u}_j - \underline{v}_j\|_{C(\bar{\Omega})}).$$

We can prove the results concerning $\underline{S}_i$ with the same arguments.

In the sequel we select the parameters $d_i \ge 0$ such that

$$d_i \ge \ell_{ii}, \quad i=1(1)m. \tag{2.15}$$

This way the mappings

$$s \to d_i s_i + f_i(x,s)$$

are isotone w.r.t. the i-th component of s.

<u>Lemma 2</u>: Let $U = [\underline{u},\bar{u}] \subset U^0$ with $\underline{u},\bar{u} \in C(\bar{\Omega})^m$ denote some fixed point of the operator T. Then $\underline{u} = \bar{u} =: u$ holds and u forms a weak solution of system (2.1).

<u>Proof</u>: By Lemma 1 we have $\bar{S}U, \underline{S}U \in C(\bar{\Omega})^m$. Using

$(L+D_i)^{-1}: L_2(\Omega) \to H_0^1(\Omega) \cap H^2(\Omega)$ and

$$U = T\,U$$

this results in $\underline{u},\bar{u} \in H_0^1(\Omega)^m \cap H^2(\Omega)^m$. Let denote $w := \bar{u} - \underline{u}$. Now, (2.3), (2.5) lead to

$$(L+D_i) w_i = \bar{S}_i U - \underline{S}_i U.$$

Definition (2.12) of the operators $\bar{S}, \underline{S}$ and (2.9),(2.15) yield

$$\bar{S}_i U \le d_i \bar{u}_i + f_i(\cdot,\bar{u}) + \sum_{j \ne i} \ell_{ij}(\bar{u}_j - \underline{u}_j)$$

$$\underline{S}_i U \ge d_i \underline{u}_i + f_i(\cdot,\bar{u}_1,\dots,\bar{u}_{i-1},\underline{u}_i,\bar{u}_{i+1},\dots,\bar{u}_m) - \sum_{j \ne i} l_{ij}(\bar{u}_j - \underline{u}_j)$$

Thus, we have

$$Lw_i+f_i(.,\bar{u}_1,..,\bar{u}_{i-1},\underline{u}_i,\bar{u}_{i+1},..,\bar{u}_m)-f_i(.,\bar{u}) \le 2\sum_{j\neq i} \ell_{ij}w_j. \qquad (2.16)$$

Taking into account the one sided Lipschitz continuity of f_i multiplying inequality by w_i and integrating by parts we obtain

$$(w_i',w_i') + \ell_i(w_i,w_i) \le 2 \sum_{j\neq i} \ell_{ij}(w_j,w_i).$$

Where $(.,.)$ denotes the scalar product in $L_2(\Omega)$. This results in $\|w_i\|_{L_2(\Omega)} = 0$ because the matrix A defined by (2.11) is assumed to be positive definite. Thus, we have $\underline{u} = \bar{u}$ and $u := \underline{u} = \bar{u}$ solves (2.1) due to the definition of the operator T.

Theorem 1: Under the assumptions made above the iteration scheme (2.8) converges monotonically to the unique solution $u \in U^o$ of system (2.1).

Proof: The monotonicity of the method results from (2.6) and $(L+D_i)^{-1} \ge 0$. With Lemma 1, $(L+D_i)^{-1}: L_2(\Omega) \to H^2(\Omega)$ and the compact embedding $H^2(\Omega) \subsetneq C(\Omega)$ we know U^k to converge to some interval $U = [\underline{u},\bar{u}]$; $\underline{u},\bar{u} \in C(\bar{\Omega})^m$ by standard arguments (compare [3] e.g.). The remaining part follows from lemma 2.

3. Monotone discretization of the iteration map

The method proposed in chapter 2 leads to a sequence of systems of linear boundary value problems (2.3) which are decoupled. Thus, they can be solved independently in difference to the problems generated in [4]. The idea of monotone discretization consists in replacing the right hand sides of (2.3) by simple ones which overestimate and underestimate, respectively, the original functions $\bar{S}_iU, \underline{S}_iU$. Here we use piecewise constant replacements. The techniques discussed in [1],[5] can be adapted to systems of boundary value problems in a similar way.

Let $\{x_j\}_{j=0}^N$ denote some given grid on Ω, i.e.

$$a = x_0 < x_1 < \dots < x_{N-1} < x_N = b .$$

We denote

$$h_j := x_j - x_{j-1},\ \Omega_j := (x_{j-1},x_j),\ j=1(1)N.$$

The mesh size h of the grid is given by

$$h := \max_{1\le j\le N} h_j.$$

By means of functions

$$\zeta_j(x) := \begin{cases} 1 , & \text{if } x \in \Omega_j \\ 0 , & \text{otherwise} \end{cases} \qquad (3.1)$$

we define $P_o := \mathrm{lin}\{\zeta_j\}_{j=1}^{N} \subset L_2(\Omega)$ as the space of piecewise constant functions. Let $\bar{p}_h, \underline{p}_h: C(\bar{\Omega}) \to P_o$ denote the following operators

$$\left.\begin{array}{l} [\bar{p}_h w](x) := \sup\limits_{y \in \Omega_j} w(y) \\ \text{and} \\ [\underline{p}_h w](x) := \inf\limits_{y \in \Omega_j} w(y) \end{array}\right\} \text{ for any } x \in \Omega_j,\ j=1(1)N. \tag{3.2}$$

Now, we replace the right hand sides $\bar{S}_i U$, $\underline{S}_i U$ in system (2.3) by $\bar{p}_h \bar{S}_i U$, $\underline{p}_h \underline{S}_i U$. Instead of (2.3) we relate functions $\bar{v}_i, \underline{v}_i$ to a given interval $U = [\underline{u}, \bar{u}]$ by

$$\begin{aligned} -\bar{v}_i'' + d_i \bar{v}_i &= \bar{p}_h \bar{S}_i U \quad \text{a.e. in } \Omega \\ \bar{v}_i(a) &= \bar{v}_i(b) = 0 \end{aligned} \tag{3.3}$$

and

$$\begin{aligned} -\underline{v}_i'' + d_i \underline{v}_i &= \underline{p}_h \underline{S}_i U \quad \text{a.e. in } \Omega \\ \underline{v}_i(a) &= \underline{v}_i(b) = 0 \ . \end{aligned}$$

Thus, an operator T_h mapping intervals $[\underline{u}, \bar{u}]$ into intervals $[\underline{v}, \bar{v}]$ can be defined by (3.3) and

$$T_h[\underline{u}, \bar{u}] := [\underline{v}, \bar{v}]. \tag{3.4}$$

Similar to algorithm 1 this leads to the following monotone iteration scheme which results in a sequence of monotonically convergent exponential splines in each component.

<u>Algorithm 2:</u>

Step 1: Select some interval $U^o = [\underline{u}^o, \bar{u}^o]$ satisfying (2.2) and

$$U^o \supset T_h\, U^o \ . \tag{3.5}$$

Step 2: Determine recursively

$$U^{k+1} := T_h\, U^k \tag{3.6}$$

for $k=0,1,\dots$.

<u>Lemma 3</u>: The mappings $\bar{p}_h \bar{S}_i$, $\underline{p}_h \underline{S}_i$: $C(\bar{\Omega})^m \times C(\bar{\Omega})^m \to L_2(\Omega)$ are continuous for any $\underline{u}, \bar{u} \in U^o$.

<u>Proof:</u> The assertion above easily follows from lemma 1 and because $\bar{p}_h, \underline{p}_h: C(\bar{\Omega}) \to L_2(\Omega)$ according to (3.2) are continuous.

Theorem 2: Algorithm 2 generates a sequence of intervals $\{U^k\}$ converging to some interval $U_h = [\underline{u}_h, \bar{u}_h]$ monotonically, i.e.

$$U_h \subset \dots \subset U^{k+1} \subset U^k \subset \dots \subset U^0$$

and

$$\lim_{k\to\infty} \|\underline{u}_h - \underline{u}^k\|_{C(\bar{\Omega})} = 0, \quad \lim_{k\to\infty} \|\bar{u}_h - \bar{u}^k\|_{C(\bar{\Omega})} = 0 .$$

Furthermore U_h contains a unique solution of problem (2.1).

Proof: Because of the monotonicity properties of the mappings $\bar{S}, \underline{S}$ and $\bar{P}_h, \underline{P}_h$ and because of $(L+D_i)^{-1} \geq 0$ we have

$$[\underline{w}, \bar{w}] \subset [\underline{u}, \bar{u}] \text{ implies } T_h[\underline{w}, \bar{w}] \subset T_h[\underline{u}, \bar{u}]. \tag{3.7}$$

Now, (3.5),(3.6) recursively result in

$$U^{k+1} \subset U^k, \; k=0,1,\dots .$$

The continuity stated in lemma 3 and the compactness arguments mentioned in the proof of theorem 1 lead to the existence of some limit interval U_h of $\{U^k\}$. Furthermore we have

$$U_h = T_h \, U_h. \tag{3.8}$$

The definitions (2.12),(3.2)-(3.4) yield $T_h U \supset T\, U$ for any interval U. Now, (3.8) results in

$$U_h \supset T\, U_h.$$

Thus, U_h can be taken as a starting interval in algorithm 1. Theorem 1 shows a unique $u \in U_h$ to exist which solves (2.1).

Next, we investigate the convergence of U_h w.r.t. $h \to 0$ on a family of refined grids (compare [1]).

Theorem 3: In the case of refined grids we have

$$\lim_{h\to+0} \|\bar{u}_h - \underline{u}_h\|_{C(\bar{\Omega})} = 0$$

for the functions $\underline{u}_h, \bar{u}_h$ forming the limit interval generated by algorithm 2.

Proof: In sequences of refined grids we can use the limit interval from the coarser grid as a starting interval for algorithm 2 on a finer grid. This results in inclusions

$$U_{h'} \subset U_h \quad \text{for} \quad h' < h .$$

With the uniform boundedness of $\{\underline{u}_h\}, \{\bar{u}_h\}$ in $H_0^1(\Omega)$ and the compact embedding $H_0^1(\Omega) \subsetneq C(\bar{\Omega})$ this leads to

$$\lim_{h\to+0}\|\underline{u}_h-\underline{u}\|_{C(\bar\Omega)}=0,\ \lim_{h\to+0}\|\bar u_h-\bar u\|_{C(\bar\Omega)}=0$$

with some $\underline{u},\bar{u}$ satisfying

$$\underline{u}_h \le \underline{u} \le \bar u \le \bar u_h \quad \text{for any } h.$$

Definition (3.2) results in

$$\lim_{h\to+0}\|\bar p_h \bar S_i U_h - \bar S_i U\|_{L_2(\Omega)} = 0 .$$

Theorem 2 and the continuity of $(L+D_i)^{-1}: L_2(\Omega)\to H^2(\Omega)$ yield

$$U = T\,U$$

with $U=[\underline{u},\bar u]$. Theorem 1 completes the proof.

Next we deal with the finite dimensional performance of algorithm 2. With (3.3),(3.4) the iteration (3.6) is equivalent to

$$-\bar u_i^{k+1\,\prime\prime} + d_i\,\bar u_i^{k+1} = \bar z_i^k,\quad \bar u_i^{k+1}(a)=\bar u_i^{k+1}(b)=0$$

and (3.9)

$$-\underline{u}_i^{k+1\,\prime\prime} + d_i\,\underline{u}_i^{k+1} = \underline{z}_i^k,\quad \underline{u}_i^{k+1}(a)=\underline{u}_i^{k+1}(b)=0,\quad i=1(1)m$$

with

$$\bar z_i^k := \bar p_h \bar S_i U^k,\quad \underline{z}_i^k := \underline{p}_h \underline{S}_i U^k.$$

Because of $\bar z_i^k,\ \underline{z}_i^k \in P_0$ we have

$$\left.\begin{aligned}\bar z_i^k(x) &\equiv \bar z_{ij}^k\\ \underline{z}_i^k(x) &\equiv \underline{z}_{ij}^k\end{aligned}\right\}\quad \text{for any } x\in\Omega_j$$

with constants $\bar z_{ij}^k,\ \underline{z}_{ij}^k \in R$, $i=1(1)m$, $j=1(1)N$.

Let functions $\varphi_{ij},\psi_{ij}: \bar\Omega \to R$ be defined by

$$\varphi_{ij}(x) := \begin{cases} \dfrac{\sinh(\sqrt{d_i}(x-x_{j-1}))}{\sinh(\sqrt{d_i}\,h_j)}, & x\in\Omega_j\\[2ex] \dfrac{\sinh(\sqrt{d_i}(x_{j+1}-x))}{\sinh(\sqrt{d_i}\,h_{j+1})}, & x\in\Omega_{j+1}\\[2ex] 0, & \text{otherwise};\ i=1(1)m,\ j=0(1)N\end{cases} \tag{3.10}$$

and

$$\psi_{ij}(x) := \begin{cases} (1-\varphi_{ij-1}(x)-\varphi_{ij}(x))/d_i, & x\in\Omega_j\\[1ex] 0, & \text{otherwise};\ i=1(1)m,\ j=1(1)N.\end{cases} \tag{3.11}$$

Now, the i-th component $\bar{u}_i^{k+1}$ of (3.9) has a representation

$$\bar{u}_i^{k+1}(x) = \sum_{j=1}^{N-1} \bar{u}_{ij}^{k+1}\varphi_{ij}(x) + \sum_{j=1}^{N} \bar{z}_{ij}^{k}\psi_{ij}(x). \tag{3.12}$$

Here the unknown coefficients $\bar{u}_{ij}^{k+1}$ can be determined from $\bar{u}_i^{k+1} \in H_0^1(\Omega) \cap H^2(\Omega)$. Because of $H^2(\Omega) \hookrightarrow C^1(\bar{\Omega})$ this leads to the conditions

$$\bar{u}_i^{k+1\,\prime}(x_j-0) = \bar{u}_i^{k+1\,\prime}(x_j+0), \quad j=1(1)N-1. \tag{3.13}$$

Using (3.10)-(3.12) system (3.13) is equivalent to the following tri-diagonal linear system

$$\begin{aligned} -a_{ij}\bar{u}_{ij-1}^{k+1}+b_{ij}\bar{u}_{ij}^{k+1}-a_{ij+1}\bar{u}_{ij+1}^{k+1} &= \bar{r}_{ij}^{k}, \quad j=1(1)N-1 \\ \bar{u}_{io}^{k+1} = \bar{u}_{iN}^{k+1} &= 0 \end{aligned} \tag{3.14}$$

with the coefficients

$$a_{ij} := 1/\sinh(\sqrt{d_i}h_j)$$

$$b_{ij} := a_{ij}\cosh(\sqrt{d_i}h_j) + a_{ij+1}\cosh(\sqrt{d_i}h_{j+1})$$

$$\bar{r}_{ij}^{k} := \frac{\cosh(\sqrt{d_i}h_j)-1}{d_i}\, a_{ij}\, \bar{z}_{ij}^{k} + \frac{\cosh(\sqrt{d_i}h_{j+1})}{d_i}\, a_{ij+1}\, \bar{z}_{ij+1}^{k}.$$

System (3.14) is symmetric and positive definite. Thus, fast factorization can be applied for the numerical solution of (3.14). For furthe details the interested reader is referred to [5]. Related systems (3.14) can be derived for the lower bounds $\underline{u}^{k+1}$ as well.

Remarks:

- We can select a sequence of functions $d_i^k \in P_o$ instead of a single constant $d_i \geq 0$. This way the convergence can be improved (compare [1],[5]).
- Further simplifications in the numerical implementation can be realized by using various kinds of estimators replacing exact minimization and maximization, respectively, needed to determine as well $\bar{S}_iU^k$, $\underline{S}_iU^k$ as the related piecewise constant bounds. It is important but rather technical to combine the method proposed here with principles given in [1].
- The principle of generating initial bounds satisfying (3.5) given in [1] can be modified to the case we consider here.

Finally, we sketch results obtained by a computer realization of algorithm 2 on IBM PC. There we used piecewise constant bounds $Q^k \supset U^k$ of the interval solutions, i.e. $Q^k = [\underline{q}^k, \overline{q}^k]$ with $\underline{q}_i^k, \overline{q}_i^k \in P_o$, $i=1(1)m$, $k=0,1,\ldots$, and we replaced $\overline{pS}_iU^k$, $\underline{pS}_iU^k$ by $\overline{pS}_iQ^k$ and $\underline{pS}_iQ^k$, respectively.

We report some results obtained by algorithm 2 applied to the following example:

$$-u_1'' = -\exp(u_1) + 0.25\sin(u_2)+\exp(x^2(1-x))-$$

$$-0.25\sin(0.2\sin(2\pi x)) - 2 + 6x \ , \ x \in (0,1)$$

$$-u_2'' = 2\sin(u_1) - 1.5\exp(u_2)-2\sin(x^2(1-x)) +$$

$$+1.5\exp(0.2\sin(2\pi x)) + 0.8\pi^2\sin(2\pi x)$$

$$u_1(0) = u_1(1) = 0 \ , \qquad u_2(0) = u_2(1) = 0 \ .$$

This problem possesses the unique solution

$$u_1(x) = x^2(1-x); \qquad u_2(x) = 0.2\sin(2\pi x).$$

We used the functions $\underline{u}_i^o \equiv -1$, $\overline{u}_i^o \equiv 1$, $i=1,2$ as initial guess which form the bounds of the starting interval U^o. The iteration was terminated by the stopping criterion

$$|\delta_k - \delta_{k-1}| \leq 10^{-3}|\delta_k|$$

with

$$\delta_k := \max_{1\leq i\leq m} \|\overline{u}_i^k - \underline{u}_i^k\|_{C(\bar{\Omega})} \ .$$

We obtained the following finalvalues $\delta = \delta_k$ for different numbers N of grid points

N	25	50	100	200
δ	0.0875	0.0439	0.0220	0.0110

The results show a linear convergence of algorithm 2 which results from using piecewise constant bounds. An improvement can be achieved by applying higher order approximations of $\overline{S}_iU^k$, $\underline{S}_iU^k$.

The following figure shows as well the obtained bounds as the known exact solutions.

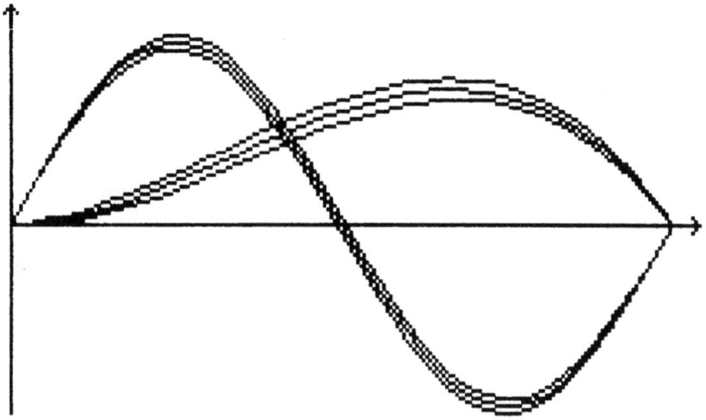

Next, we consider the problem

$$-u_1'' = -\alpha(\sinh u_1 - \operatorname{sgn}((x-0.55)(x-0.75))) + \sin(u_2) \qquad x \in (0,1)$$

$$-u_2'' = 10u_1 - u_2 + 1$$

$$u_1(0) = u_1(1) = 0 \;,\; u_2(0) = u_2(1) = 0 \;.$$

Here $\alpha \in R$, $\alpha > 0$ denotes some parameter. Solution u_1 has interiour layers for large α near 0.55 and 0.75 as well as boundary layers. We obtained with $\alpha = 100$ the results

N	20	40	80
δ	0.4406	0.2283	0.1171

and the following figure shows the computed bounds in the case N = 80.

References

[1] Al-Zanaidi,M.; Großmann,C.: Monotone iteration discretization for BVP's.
Computing (1988).

[2] Amann,H.: Invariant sets and existence theorems for semilinear parabolic and elliptic systems.
J.Math.Anal.Appl. 65(1978), 432-467.

[3] Carl,S.: A monotone iterative scheme for nonlinear reaction-diffusion systems having nonmonotone reaction terms.
J.Math.Anal.Appl. 134(1988)1, 81-93.

[4] Carl,S.: An enclosing theorem and a monotone iterative scheme for elliptic systems having nonmonotone nonlinearities.
(to appear in ZAMM).

[5] Großmann,C.: Monotone discretization of two-point boundary value problems and related numerical methods.
In: Adams,E.; Ansorge,R.; Großmann,C.; Roos,H.-G. (eds.): Discretization in differential equations and enclosures.
Akademie-Verlag, Berlin, 1987, 99-122.

[6] Ladde,G.S.; Lakshmikantham,V.; Vatsala,A.S.:
Existence of coupled quasisolutions of systems of nonlinear reaction-diffusion equations.
J.Math.Anal.Appl. 108(1985), 249-266.

[7] Sattinger,D.H.: Monotone methods in nonlinear elliptic and parabolic boundary value problems.
Indiana Univ.Math.J. 21(1972), 979-1000.

ALGORITHMS FOR SHAPE-PRESERVING INTERPOLATION

PAOLO COSTANTINI

Abstract. *The scope of this paper is to give a short description of some algorithms for the construction of shape-preserving surfaces interpolating arbitrary sets of data on rectangular grids. It consists of a brief survey of methods appeared in the last years, and in a more detailed explanation of an algorithm recently developed by F.Fontanella and the author. Graphical examples and the outline of the related FORTRAN code are reported.*

Key words. *splines, Bernstein-Bezier polynomials, Bezier nets, tensor-product spaces, blending functions methods, shape-preserving interpolation, CAGD.*

AMS (MOS) subject classifications. 65D05, 65D07.

1. Introduction. The amount of papers concerning surface fitting (or, more generally, multivariate approximation) and one dimensional (1-D) shape-preserving interpolation (SPI) has rapidly grown during the last years, because of their theoretical interest and of their applications in CAGD (the interested reader is referred to [8] and [16] for a basic text and an excellent bibliography on multivariate approximation respectively, to the references quoted in [11], [24] for 1-D, SPI and to [5] and its references for an idea of their applications in CADG). However, two-dimensional (2-D) SPI, which lies in the intersection of the two fields and is of great interst in CAGD, is still not well understood, and only few methods ([1], [15], [4], [27], [2], [14]) are, at least in our knowledge, available.

For obvious reasons of space, it is impossible to survey all these papers, and we are going to restrict our attention to the interpolation of gridded data, giving a description of the methods proposed in [15], [4], [2], and a more detailed account of the papers [14], [13]. However, it is worthwhile to mention here the very interesting approach of F.I.Utreras ([25], [26], [27]), in which positive or monotone interpolating splines are obtained as the solutions of constrained minimization problems in suitable Sobolev spaces. This approach (which was then studied in [23] from a more general point of view), is global in nature, and, although no comparison has been published, it may be argued that it works slower than ad-hoc techniques when applied to gridded data. On the other hand it has the very important features of being appliable to arbitrary data sets (in particular scattered) and to the interpolation in more variables, and of allowing general results on the approximation properties.

2. Lagrange type interpolation in fixed spaces. Let $x_0<x_1<...<x_N$, $y_0<y_1<...<y_M$ be two ordered sets of real numbers, with N , $M \geq 2$, and let us consider the rectangles $R = [x_0,x_N]\times[y_0,y_M]$; $R_{i,j} = [x_i,x_{i+1}]\times[y_j,y_{j+1}]$, $i=0,1,...,N-1$, $j=0,1,...,M-1$,
and the gridded data

(2.1) $\left\{ (x_i,y_j,f_{i,j}) \ ; \ i=0,1,...,N \ , \ j=0,1,...,M \right\}$.

We then introduce the following terms

$H_i=x_{i+1}-x_i \ ; \ \Delta_{i,j}=(f_{i+1,j}-f_{i,j})/H_i \ , \ i=0,1,...,N-1 \ ;$

$Z_j=y_{j+1}-y_j \ ; \ \theta_{i,j}=(f_{i,j+1}-f_{i,j})/Z_j \ , \ j=0,1,...,M-1 \ ,$

and define the shape properties of the data starting from their sections along the grid lines (these definitions are stated for the x variable; the y case is straightforward). We will say that the data are *increasing* (*I*) (*decreasing* (*D*)) *and/or convex* (*Cnv*) (*concave* (*Cnc*)) *with respect to* x *in the set* $[x_i,x_{i+1}]\times\{y_j\}$ *if*

(2.2) $\Delta_{i,j}\geq 0 \quad (\Delta_{i,j}<0)$ *and/or* $\Delta_{i-1,j}\leq\Delta_{i,j} \quad (\Delta_{i-1,j}>\Delta_{i,j})$.

(Note that the equalities have no practical influences and are included in I and Cnv cases for notational convenience). Starting from (2.2) we will say that the data are *I* (*D*) *and/or Cnv* (*Cnc*) *in* $R_{i,j}$ *if they are I* (*D*) *and/or Cnv* (*Cnc*) *in* $[x_i,x_{i+1}]\times\{y_j\}$ *and in* $[x_i,x_{i+1}]\times\{y_{j+1}\}$. Moreover we will say that the data are simply *I* (*D*) *and/or Cnv* (*Cnc*) *if they are I* (*D*) *and/or Cnv* (*Cnc*) *for all the rectanles* $R_{i,j}$. With the term SP *interpolating function* we mean an interpolant which has *the same shape of the data in the same regions*. Further, we will use the terms *I* (*D*) *and/or Cnv* (*Cnc*) *interpolant* when the function is *I* (*D*) *and/or Cnv* (*Cnc*) *in* R , and the terms *co-monotone* or *co-convex interpolant* when *only the I/D or Cnv/Cnc properties of the data are checked and the function is required to be only I/D or Cnc/Cnc in the same regions* (in other words it is required to be SP in a restricted sense).

2.1. A blending function approach. Since, as previously said in the introduction, many efficient algorithms are available for 1-D, SP problems, it seems an obvious approach to use one of them for the interpolation along grid lines, and a suitable method for connecting toghether these curves in such a way of obtaining a sufficiently smooth function in the whole domain. This idea has been followed by Dodd, Mc Allister and Roulier ([15]), using the 1-D quadratic spline of [22] for the grid lines and a blending function technique for the second step. We start the discussion giving a short explanation of the blending function method used in [15]. For further details and further references, the reader is referred to [5], [3] and [18]. For simplicity we consider the unit square $S=[0,1]\times[0,1]$; the extention to generic $R_{i,j}$ is straightforward. Let f and its continous first order partial derivatives be given along the boundaries of S , and suppose that the second mixed partial derivatives are given at the four corners of S . We introduce the cardinal basis functions for cubic Hermite interpolation

$\phi_1(t)=(t-1)^2(2t+1) \ ; \ \phi_2(t)=t^2(-2t+3) \ ;$

$\phi_3(t)=(t-1)^2t \ ; \ \phi_4(t)=t^2(t-1) \ ,$

the operators

$P_1 \ : \ P_1f(x,y) = f(0,y)\phi_1(x) + D_xf(0,y)\phi_3(x) + f(1,y)\phi_2(x) + D_xf(1,y)\phi_4(y)$

$P_2 \ : \ P_2f(x,y) = f(x,0)\phi_1(y) + D_yf(x,0)\phi_3(y) + f(x,1)\phi_2(y) + D_yf(x,1)\phi_4(y)$

and the blending function $(P_1+P_2-P_1P_2)f$ given by

(2.3) $(P_1+P_2-P_1P_2)f(x,y) = P_1f(x,y) + P_2f(x,y) - f(0,0)\phi_1(x)\phi_1(y)-$
$f(1,0)\phi_2(x)\phi_1(y)-f(0,1)\phi_1(x)\phi_2(y)-f(1,1)\phi_2(x)\phi_2(y)-D_xf(0,0)\phi_3(x)\phi_1(y)-$
$D_xf(1,0)\phi_4(x)\phi_1(y)-D_xf(0,1)\phi_3(x)\phi_2(y)-D_xf(1,1)\phi_4(x)\phi_2(y)-$
$D_yf(0,0)\phi_1(x)\phi_3(y)-D_yf(1,0)\phi_2(x)\phi_3(y)-D_yf(0,1)\phi_1(x)\phi_4(y)-$
$D_yf(1,1)\phi_2(x)\phi_4(y)-D_{xy}f(0,0)\phi_3(x)\phi_3(y)-D_{xy}f(1,0)\phi_4(x)\phi_3(y)-$
$D_{xy}f(0,1)\phi_3(x)\phi_4(y)-D_{xy}f(1,1)\phi_4(x)\phi_4(y)$

It is simple to prove that P_1f is a $C^1(S)$ function interpolating f and its x first partial derivatives along the left and the right edge of S (similar features are performed by P_2f) and that $(P_1+P_2-P_1P_2)f$ interpolates the function and the normal derivatives at the boundary of S , if $D_{xy}f=D_{yx}f$ at the four corners of S. In this case P_1P_2f is the classical tensor product Hermite interpolant. Note that the blending function also interpolates the four mixed partial derivatives at the corners. It is evident that this scheme is adaptable to any $R_{i,j}$ and that a $C^1(R)$ surface is obtainable specifying both the function and the normal derivatives along the grid lines. The shape preserving method of Dodd, Mc Allister and Roulier consists in using the well-known quadratic 1-D approach for generating the curves and the derivatives on the meshes. A detailed description of such method is beyond the scope of this paper, and the reader is thus referred to [21], [22]. We only recall that, given the sequences of knots $\tau_0,\tau_1,\ldots,\tau_N$ and the values $\eta_0,\eta_1,\ldots,\eta_N$, $\delta_0,\delta_1,\ldots,\delta_N$ it is possible to construct a C^1 quadratic polynomial which interpolates (in Hermite sense) the values (τ_i,η_i,δ_i) , $(\tau_{i+1},\eta_{i+1},\delta_{i+1})$, $i=0,1,\ldots,N-1$ and which locally preserves the shape of the data, by adding one or two suitable knots for each interval $[\tau_i,\tau_{i+1}]$. In [22] is described how to choose the abscissas and the ordinates of these knots; moreover, since the δ_i are not always available, and generally a Lagrange-type interpolation is required, a method for computing derivatives consistently with the shape of the η_i is provided. The 2-D algorithm can be summarized as follow:

Step 1. *For any y_j compute the quadratic spline $s_j(x)$ (Lagrange) interpolating the data $(x_i,f_{i,j})$, $i=0,1,\ldots,N$. Similarly, for any x_i , compute $s_i(y)$ interpolating the data $(y_j,f_{i,j})$, $j=0,1,\ldots,M$.*

Step 2. *For any y_j compute the quadratic spline $\sigma_j(x)$ (Hermite) interpolating the data $(x_i,s_i'(y_j),0)$, $i=0,1,\ldots,N$ ($\sigma_j(x)$ represents the normal derivative along the line $y=y_j$) . Similarly, for any x_i , compute $\sigma_i(y)$ interpolating the data $(y_j,s_j'(x_i),0)$, $j=0,1,\ldots M$.*

Step 3. *Construct the $C^1(R)$ surface using, for any $R_{i,j}$, the blending functions given by analogous of formulas (2.3) .*

(Although not necessary, the choice $\sigma_i'(y_j)=\sigma_j'(x_i)=0$ in step 2 is widely motivated in [15] by heuristic considerations, and avoids unwanted "twists" in the interpolant). We note that the surface is SP along the mesh lines (at step 2 co-monotone partial derivatives are produced), but this may not happen inside the subrectangles for some pathological example. However, since this method produces, in general, visually pleasing plots (see [15, pp. 74-77]) and is appliable to any set of gridded data, it seems to be one of the most reliable for CAD applications.

2.2. A natural approach. In this subsection we consider the method of Carlson and Fritsch (developed in the sequence of papers [6], [18], [7]), of Beatson and Ziegler ([4]) and the more recent one of Asaturyan and Unsworth ([2]). The constructive approaches are basically the same, and represent the generalization to the 2-D case of the most used 1-D method. This justifies the adjective "natural" in the title. We start following the recent report [7], which will be more emphasized than the other papers because of its similarities with the author's researches.

Let

$$SX_3^1 = \left\{ s\in C^1[x_0,x_N] : s(x)=p(x) \ , \ p\in PX_3 \ all \ x\in[x_i,x_{i+1}] \ , \ i=0,1,\dots,N-1 \right\}$$

$$SY_3^1 = \left\{ s\in C^1[y_0,y_M] : s(y)=p(y) \ , \ p\in PY_3 \ all \ y\in[y_j,y_{j+1}] \ , \ j=0,1,\dots,M-1 \right\}$$

(PX_3^1 and PY_3^1 stand for the sets of cubic polynomials in x and y) be usual spaces of cubic C^1 splines, and let

$$SXY_{3,3}^1 = SX_3^1 \otimes SY_3^1$$

be the space of tensor product, C^1, cubic spline. It is immediate to verify that $dim(SXY_{3,3}^1) = (2N+2)(2M+2)$, and that, for any $s\in SXY_{3,3}^1$, $s|_{R_{i,j}} = p_{i,j}\in PX_3\otimes PY_3$. Moreover, any s is uniquely defined by the $4(2N+2)(2M+2)$ values $s(x_i,y_j)$, $D_x s(x_i,y_j)$, $D_y s(x_i,y_j)$, $D_{x,y} s(x_i,y_j)$, and, since we have from (2.1), $s(x_i,y_j)=f_{i,j}$, it remains to find a way to determine the others parameters. The Fritsch and Carlson's sequence of papers is concerned with monotone interpolation. For simplicity we will suppose that the data are increasing with respect to x. The constraints for the other cases are easily obtainable and are, in any case, reported in [7]. Let

$$E = \left\{ (u,v)\in\mathbf{R}^2 : u\geq 0 \ , \ v\geq 0 \ , \ u+v\leq 3 \right\} \cup \left\{ (u,v)\in\mathbf{R}^2 : \left(\frac{u+v-4}{2}\right)^2 + \left(\frac{u-v}{2}\right)^2 \leq 1 \right\}$$

be the set given by the union of a triangle of vertices (0,0), (0,3), (3,0) and the ellipse centered at (2,2) with principal axes at $u-v=0$ and at $u+v=4$. For details see [17]. In [6] is reported the following result.

Theorem 1. *Let the data (2.1) be increasing in* x. *Then the interpolating* $s\in SXY_{3,3}^1$ *is increasing in* x *if, and only if,*

(2.4,x) $s(x_i,y)\leq s(x_{i+1},y)$; $D_x s(x_i,y)=D_x s(x_{i+1},y)=0$ *if* $s(x_i,y)=s(x_{i+1},y)$;

$(D_x s(x_i,y),D_x s(x_{i+1},y)) \in E\cdot(s(x_i,y)-s(x_{i+1},y))/H_i$; $\forall \ y\in[x_0,x_N]$, $i=0,1,\dots,N-1$.

The inequalities (2.4,x) are equivalent to require that $D_x s(x,y)\geq 0$ for any $(x,y)\in R$; unfortunately, as also remarked by the authors, they are extremely dificult to use in practice. Hence, it is necessary to develop a set of simpler sufficient conditions.

Theorem 2. *Let the data (2.1) be increasing in* x. *Then the interpolating* $s\in SXY_{3,3}^1$ *is increasing in* x *if, for all admissible indices*

(2.5,x) $0\leq D_x s(x_i,y_j), D_x s(x_{i+1},y_j)\leq 3\Delta_{i,j}$;

(2.6,x) $-b_{i,j}/z_j \leq (D_x s(x_{i+1},y_j)-D_x s(x_i,y_j)) \leq b_{i,j}/z_{j-1}$, *where*

$b_{i,j} = H_i(3\Delta_{i,j}-max\{D_x s(x_i,y_j),D_x s(x_{i+1},y_j)\})\geq 0$;

$$(2.7,x)\quad -3D_x s(x_i,y_j)/z_j \le D_{xy} s(x_i,y_j) \le 3(D_{i,j}-D_x s(x_i,y_j)/z_j) \; ,$$

$$3(C_{i,j}+D_x s(x_i,y_j)/z_{j-1}) \le D_{xy} s(x_i,y_j) \le 3(D_{i,j}-D_x s(x_i,y_j)/z_j) \; ,$$

$$3(C_{i,j}+D_x s(x_i,y_j)/z_{j-1}) \le D_{xy} s(x_i,y_j) \le 3D_x s(x_i,y_j)/z_{j-1} \; , \text{ where}$$

$$C_{i,j} = (D_x s(x_{i+1},y_j)-D_x s(x_i,y_j))/H_i - 3\Delta_{i,j}/z_{j-1} \; ,$$

$$D_{i,j} = (D_x s(x_{i+1},y_j)-D_x s(x_i,y_j))/H_i + 3\Delta_{i,j}/z_j \; .$$

If we also consider monotonicity with respect to y , then inequalities (2.5,y)-(2.7,y), analogous (2.5,x)-(2.7,x), can be obtained. We refer to (2.✱) for matching both the cases. Note that, if (2.6) are satisfied, then (2.7) will define intervals containing zero. In other words, even if (2.7,x) and (2.7,y) represent bounds for the same quantity $-D_{xy}s(x_i,y_j)$ - such bounds are not incompatible. This observation is crucial for understanding the improvement that has been made in the last version of the method. Note also that a solution does always exist, since the zero partial derivatives satisfy the constraints. In the case of data increasing in both the variables, the algorithm suggested in [7] goes as follow.

Step 1. *Verify that the data are increasing.*

Step 2. *Compute initial values for* $D_x s(x_i,y_j)$ *and* $D_y s(x_i,y_j)$ *which satisfy* (2.5).

Step 3. *Construct an interval of acceptable values for* $D_{xy} s(x_i,y_j)$ *changing the first derivatives in such a way that* (2.6) *are satisfied.*

Step 4. *Compute values for* $D_{xy} s(x_i,y_j)$ *which satisfy* (2.7).

The basic concept in this algorithm is to make the changes in the derivatives as small as possible; the proposed version (for details see [7]) produces an I surface (which is independent of the order of data processing) after a finite number of steps. The options allowed concern the possibility of imposing monotonicity in only one variable (which, of course, reduces the computational time) and of introducing boundary conditions on the derivatives.

In the remaining of this section, we briefly sketch the methods proposed in [4] and in [2]. For introducing the Beatson and Ziegler's approach, we need a triangularization $T_{i,j}$ of the rectangle $R_{i,j}$ as shown in fig. 1 , and we will consider the space of piecewise quadratic C^1 functions

$$\overline{Q} = \left\{ p \in C^1(R_{i,j}) : p|_t \in P_2(\mathbb{R}^2) \;\; \text{for any triangle} \;\; t \in T_{i,j} \right\}$$

($P_2(\mathbb{R}^2)$ stands for the set of polynomial of total degree less or equal to 2). It has been proved in [4] that $dim(\overline{Q})=15$, and that a dual basis is given by the functionals which evaluate the function and its first partials at the points C_r , $r=1,2,3,4$ plus any three of the normal derivatives at $(C_r+C_{r+1})/2$, $r=1,2,3,4$, $C_5=C_1$. Beatson and Ziegler have decided to use the following simpler 12-dimensional space

$$Q_{i,j} = \left\{ p \in \overline{Q} : \frac{\partial p}{\partial n}\Big(\frac{C_r+C_{r+1}}{2}\Big) = \frac{1}{2}\Big(\frac{\partial p}{\partial n}\Big(\frac{C_r}{2}\Big) + \frac{\partial p}{\partial n}\Big(\frac{C_{r+1}}{2}\Big)\Big) \;\; i=1,2,3,4 \right\} ,$$

whose elements are expressed using the values of the function and of the first partials at the corners. Once again, a spline surface $s=s(x,y)$ interpolating (2.1) is obtained patching toghether functions of $Q_{i,j}$ for any $R_{i,j}$, and the C^1 continuity is obtained in R assigning the same partials at (x_i,y_j) for the four quadratics sharing the same corner. The use of quadratics simplifies the developement of monotonicity constraints (it suffices to impose that the partials derivatives have the same sign at the vertices of the triangle $T_{i,j}$), which are not reported here for sake of brevity. The reader is referred to [4] or to the survey paper [26] for details. We limit ourselves to mention the the facts that the conditions involving the x and the y are separated, and that the algorithm terminates after a finite number of steps. One of the most actractive features of the Beatson and Ziegler's approach is that their interpolants have good approximation properties, while, at least in our knowledge, no error estimates are available for the other methods. Supposing for simplicity that $H_0=H_1= \dots =H_N$, $Z_0=Z_1= \dots =Z_M$, we have

Theorem 3. *Let* $f \in C^3(R)$ *be monotone in* x *and in* y , *let* $L=max(H_0, Z_0)$ *and let* s *be the interpolating monotone quadratic spline. Then there exist* C_1 , C_2 *independent of* f *and* L *such that*

$$\|Df-Ds\|_\infty \le C_2 L \|D^3 f\|_\infty \quad ; \quad \|f-s\|_\infty \le C_1 L^2 \|D^3 f\|_\infty \ ,$$

where

$$\|D^3 f\|_\infty = \max_{p+q=3}\left\{\|\partial^3 f/\partial x^p \partial y^q\|_\infty\right\} \quad ; \quad \|Df\|_\infty = max\left\{ \|D_x f\|_\infty , \|D_y f\|_\infty \right\} .$$

Also in the method of Asaturyan and Unsworth ([2]) quadratic spline s are considered for the interpolation of monotonic data, but the subdivision of the initial grid is obtained taking four subrectangles (see fig. 2) for each $R_{i,j}$. It is possible to prove ([2]) that any quadratic, C^1 spline $s_{i,j}$ on this partition of $R_{i,j}$ is obtainable by the values of $s_{i,j}$, $D_x s_{i,j}$, $D_y s_{i,j}$, $D_{x,y} s_{i,j}$ at the four corners of $R_{i,j}$. As a consequence, the surface s , interpolating the data (2.1) and obtained patching toghether the $s_{i,j}$, is $C^1(R)$ because the quadratic pieces share the same derivatives along the common edges of adjacent rectangles. A set of sufficient monotonicity constraints is obtained viewing the quadratic splines for each subrectangle $R_{i,j}$ as expressed in terms of their Bezier nets (see [2] for details and [5] for a general introduction to the Bernstein-Bezier nets), and, once again, an algorithm is developed for constructing, in a finite number of steps, a set of derivatives' values satisfying the constraints.

3. Hermite type interpolation in variable spaces. In this section we will discuss the 2-D method proposed in [14], [13], which is, in turn, based on the sequence of 1-D papers [10], [11], [12]. Before starting, it is convenient to summarize some aspects of the aproaches discussed in sect. 2, in order to understand better the present one.

From the point of view of the obtainable results the techniques of sect. 2.1. and of sect. 2.2. are quite different. The first one produces visually pleasing plots for arbitrary sets of data but does not guarentee that the surface is shape preserving inside the rectangles (although this seems to happen very often in practice); the others produce surfaces which are guarenteed to preserve the shape of the data, but are appliable only to monotone sets of points (the co-monotone case seems to be still under study, see [7, p. 10] and [2, p. 17]). On the contrary, from the constructive point of view, they basically agree, because they treat a Lagrange-type interpolation problem, and the surfaces are constructed in the following steps: a) to fix a set of functions in which the interpolants are sought and whose elements are uniquely defined assigning some partial derivatives at the knots; b) to find suitable values for the needed derivatives (eventually imposing that a set of constraints is fulfilled). Note, in passing, that the popular local method of Akima [1] fits in this scheme.

The approach of [14] is different (in some sense symmetric) because it is stated for Hermite interpolation problems (although, as we will see later, it produces a method appliable also to Lagrange problems), and the interpolants are formed using variable polynomial spaces, depending on two parameters - namely the degrees in x and in y of their elements. In this case the "variables" and the "costants" of steps a) and b) are reversed. We start recalling the definitions of sect. 2., and, for any rectangle $R_{i,j}$, arbitrary but fixed, we introduce the integers $n_{i,j}$, $m_{i,j} \geq 3$ and the quantities

$$h_i = H_i/n_{i,j} \;;\; z_j = Z_j/m_{i,j} \;;$$

$$t_{i,0} = x_i \;;\; t_{i,1} = x_i + h_i \;;\; t_{i,2} = x_{i+1} - h_i \;;\; t_{i,3} = x_{i+1} \;;$$

$$u_{j,0} = y_j \;;\; u_{j,1} = y_j + z_j \;;\; u_{j,2} = y_{j+1} - z_j \;;\; u_{j,3} = y_{j+1} \;,$$

wich we use to define the following functions

$$\ell_i^{(1)}(x) = \begin{cases} 1 & \text{for } x \in [t_{i,0}\,,\, t_{i,1}) \\ (x-t_{i,1})/((n_{i,j}-2)h_i) & \text{for } x \in [t_{i,1}\,,\, t_{i,2}) \\ 0 & \text{for } x \in [t_{i,2}\,,\, t_{i,3}] \end{cases} \tag{3.1}$$

$$\ell_i^{(2)}(x) = \begin{cases} 0 & \text{for } x \in [t_{i,0}\,,\, t_{i,1}) \\ (x-t_{i,1})/((n_{i,j}-2)h_i) & \text{for } x \in [t_{i,1}\,,\, t_{i,2}) \\ 1 & \text{for } x \in [t_{i,2}\,,\, t_{i,3}] \end{cases}$$

$$\ell_i^{(3)}(x) = \begin{cases} (x-t_{i,0}) & \text{for } x \in [t_{i,0}\,,\, t_{i,1}) \\ h_i-(x-t_{i,1})/(n_{i,j}-2) & \text{for } x \in [t_{i,1}\,;\, t_{i,2}) \\ 0 & \text{for } x \in [t_{i,2}\,,\, t_{i,3}] \end{cases}$$

$$\ell_i^{(4)}(x) = \begin{cases} 0 & \text{for } x \in [t_{i,0}\,,\, t_{i,1}) \\ (x-t_{i,1})/(n_{i,j}-2\,) & \text{for } x \in [t_{i,1}\,,\, t_{i,2}) \\ (x-t_{i,3}) & \text{for } x \in [t_{i,2}\,,\, t_{i,3}] \end{cases}$$

$$\ell_j^{(1)}(y) = \begin{cases} 1 & \text{for } y \in [u_{j,0}\,,\, u_{j,1}) \\ 1-(y-u_{j,1})/((m_{i,j}-2)z_j) & \text{for } y \in [u_{j,1}\,,\, u_{j,2}) \\ 0 & \text{for } y \in [u_{j,2}\,,\, u_{j,3}] \end{cases}$$

$$\ell_j^{(2)}(y) = \begin{cases} 0 & \text{for } y \in [u_{j,0}\,,\, u_{j,1}) \\ (y-u_{j,1})/((m_{i,j}-2)z_j) & \text{for } y \in [u_{j,1}\,,\, u_{j,2}) \\ 1 & \text{for } y \in [u_{j,2}\,,\, u_{j,3}] \end{cases}$$

$$\ell_j^{(3)}(y) = \begin{cases} (y-u_{j,0}) & \text{for } y \in [u_{j,0}\,,\, u_{j,1}) \\ z_j-(y-u_{j,1})/(m_{i,j}-2) & \text{for } y \in [u_{j,1}\,,\, u_{j,2}) \\ 0 & \text{for } y \in [u_{j,2}\,,\, u_{j,3}] \end{cases}$$

$$\ell_j^{(4)}(y) = \begin{cases} 0 & \text{for } y \in [u_{j,0}\,,\, u_{j,1}) \\ (y-u_{j,1})/(m_{i,j}-2) & \text{for } y \in [u_{j,1}\,,\, u_{j,2}) \\ (y-u_{j,3}) & \text{for } y \in [u_{j,2}\,,\, u_{j,3}] \end{cases}$$

Combining the functions of (3.1)

(3.2) $\ell_{i,j}^{(q,r)}(x,y) = \ell_i^{(q)}(x)\cdot\ell_j^{(r)}(y)$; $q=1,2,3,4$; $r=1,2,3,4$,

it is possible to define the following 16-dimensional linear spaces

(3.3) $L_{n_{i,j},m_{i,j}} = span\left\{\ell_{i,j}^{(q,r)}\ ;\ q=1,2,3,4\ ;\ r=1,2,3,4\right\}$.

We introduce the Bernstein operators $B_{n_{i,j},m_{i,j}} : C(R_{i,j}) \mapsto C(R_{i,j})$ defined by

(3.4) $$(B_{n_{i,j},m_{i,j}}f)(x,y) = H_i^{-n_{i,j}} Z_j^{-m_{i,j}} \sum_{\nu=0}^{n_{i,j}} \sum_{\mu=0}^{m_{i,j}} \binom{n_{i,j}}{\nu}\binom{m_{i,j}}{\mu}\cdot$$

$$f(x_i+\nu h_i, y_j+\mu z_j)(x-x_i)^{\nu}(x_{i+1}-x)^{n_{i,j}-\nu}(y-y_j)^{\mu}(y_{j+1})^{m_{i,j}-\mu}$$

for all $(x,y)\in R_{i,j}$,

and define the space

(3.5) $$BP_{n_{i,j},m_{i,j}} = Range\left(B_{n_{i,j},m_{i,j}}\Big|_{L_{n_{i,j},m_{i,j}}}\right)$$

It is possible to prove that $L_{n_{i,j},m_{i,j}}$ and $BP_{n_{i,j},m_{i,j}}$ are isomorphic since $B_{n_{i,j},m_{i,j}}$ is an invertible map from $L_{n_{i,j},m_{i,j}}$ to $BP_{n_{i,j},m_{i,j}}$. It is immediate from (3.4)-(3.5) and from the properties of Bernstein polynomials (see [20]) that, for any $b=B_{n_{i,j},m_{i,j}}\ell\in BP_{n_{i,j},m_{i,j}}$,

$b(x_q,y_r) = \ell(x_q,y_r)$; $D_xb(x_q,y_r) = D_x\ell(x_q,y_r)$; $D_yb(x_q,y_r) = D_y\ell(x_q,y_r)$;

$D_{x,y}b(x_q,y_r) = D_{x,y}\ell(x_q,y_r)$; $q=i,i+1$; $r=j,j+1$,

and that any $\ell\in L_{n_{i,j},m_{i,j}}$ (and consequently any $b\in BP_{n_{i,j},m_{i,j}}$) is uniquely defined giving the values of ℓ , $D_x\ell$, $D_y\ell$, $D_{xy}\ell$ at the four corners of $R_{i,j}$. Note that $\ell\in L_{n_{i,j},m_{i,j}}$ is the Bernstein-Bezier net of $b=B_{n_{i,j},m_{i,j}}\ell$ and that ℓ tends to be bilinear as $n_{i,j}$ and $m_{i,j}$ increase. Now suppose that, in addition to (2.1), the data

(3.6) $fx_{i,j}$; $fy_{i,j}$; $fxy_{i,j}$; $i=0,1,\dots,N$, $j=0,1,\dots,M$,

are given. We will say that *they are I (D) and/or Cnv (Cnc) with respect to* x *in* $[x_i,x_{i+1}]\times\{y_j\}$ (*or, in other words, along the edge* $y=u_0$) *if*

$\Delta_{i,j}>0$; $fx_{i,j}>0$; $fx_{i+1,j}>0$ ($\Delta_{i,j}<0$; $fx_{i,j}<0$; $fx_{i+1,j}<0$)

and/or

$fx_{i,j}<\Delta_{i,j}<fx_{i+1,j}$; ($fx_{i,j}>\Delta_{i,j}>fx_{i+1,j}$) ,

and we follow the scheme of sect. 2. for the definitions in the other domains. (Note that, in this case, we deal with strict inequalities). We are interested in the shape-preserving properties of the polynomial $b_{i,j}\in BP_{n_{i,j},m_{i,j}}$ such that

$b(x_q,y_r) = f_{q,r}$; $D_xb(x_q,y_r) = fx_{q,r}$; $D_yb(x_q,y_r) = fy_{q,r}$;

$D_{x,y}b(x_q,y_r) = fxy_{q,r}$; $q=i,i+1$; $r=j,j+1$,

which is clearly expressible as

(3.7) $$\begin{aligned} b_{i,j} = {} & f_{i,j}b_{i,j}^{(1,1)} + f_{i+1,j}b_{i,j}^{(2,1)} + f_{i,j+j}b_{i,j}^{(1,2)} + f_{i+1,j+1}b_{i,j}^{(2,2)} + \\ & fx_{i,j}b_{i,j}^{(3,1)} + fx_{i+1,j}b_{i,j}^{(4,1)} + fx_{i,j+j}b_{i,j}^{(3,2)} + fx_{i+1,j+1}b_{i,j}^{(4,2)} + \\ & fy_{i,j}b_{i,j}^{(1,3)} + fy_{i+1,j}b_{i,j}^{(2,3)} + fy_{i,j+j}b_{i,j}^{(1,4)} + fy_{i+1,j+1}b_{i,j}^{(2,4)} + \\ & fxy_{i,j}b_{i,j}^{(3,3)} + fxy_{i+1,j}b_{i,j}^{(4,3)} + fxy_{i,j+j}b_{i,j}^{(3,4)} + fxy_{i+1,j+1}b_{i,j}^{(4,4)} , \end{aligned}$$

where the functions $b_{i,j}^{(q,r)}$ span the space $BP_{n_{i,j},m_{i,j}}$ and are given by $b_{i,j}^{(q,r)}=B_{n_{i,j},m_{i,j}}\ell_{i,j}^{(q,r)}$;

q , r = 1,2,3,4 . For this scope, we introduce the following functions

$$\rho(\tau) = \frac{\frac{\tau}{\tau-2} \cdot \sum_{\nu=1}^{\tau-1} \binom{\tau-1}{\nu}}{\frac{2}{\tau-2} \cdot \sum_{\nu=1}^{\tau-1} \binom{\tau-1}{\nu} - 2} \quad , \quad \tau \geq 3 \quad , \quad \tau \text{ integer}$$

$$\gamma_1(\xi_0,\xi_1,\eta;\tau)=\rho(\tau)\xi_0+[\tau-\rho(\tau)]\xi_1-\rho(\tau)\tau\eta \ ;$$

$$\gamma_2(\xi_0,\xi_1,\eta;\tau)=\rho(\tau)\xi_1+[\tau-\rho(\tau)]\xi_0-\rho(\tau)\tau\eta \ ;$$

$$\gamma_3(\xi_0,\xi_1,\eta;\tau)=\xi_0+(\tau-1)\xi_1-\tau\eta \ ;$$

$$\gamma_4(\xi_0,\xi_1,\eta;\tau)=\xi_1+(\tau-1)\xi_0-\tau\eta \ ,$$

and the constraints

(3.8) $\eta>0$; $\xi_0>0$; $\xi_1>0$; $\gamma_1(\xi_0,\xi_1,\eta;\tau)<0$; $\gamma_2(\xi_0,\xi_1,\eta;\tau)<0$ *(I constraints)*

(3.9) $\xi_0<\eta<\xi_1$; $\gamma_3(\xi_0,\xi_1,\eta;\tau)>0$; $\gamma_4(\xi_0,\xi_1,\eta;\tau)<0$ *(Cnv constraints)* ,

and recall from [14] the following result.

Theorem 4. *Let* $b_{i,j}$ *be given by* (3.7), *and let* $\ell_{i,j} = B^{-1}_{n_{i,j},m_{i,j}} b_{i,j}$. *Then* $b_{i,j}$ *is I (D) and/or Cnv (Cnc) in* x *along the edge* $y=u_0$ / *along the edge* $y=u_3$ / *inside the rectangle* $R_{i,j}$ *if, setting* $\xi_0=\ell_{i,j}(x_i,u_r)$; $\xi_1=\ell_{i,j}(x_{i+1},u_r)$; $\eta=(\xi_1-\xi_0)/h_i$; $\tau=n_i$, *the inequalities of* (3.8) *and/or* (3.9) *are fulfilled for* $r=0$ / $r=3$ / $r=0,1,2,3$.

$b_{i,j}$ *is I (D) and/or Cnv (Cnc) in* y *along the edge* $x=t_0$ / *along the edge* $x=t_3$ / *inside the rectangle* $R_{i,j}$ *if, setting* $\xi_0=\ell_{i,j}(t_q,y_j)$; $\xi_1=\ell_{i,j}(t_q,y_{j+1})$; $\eta=(\xi_1-\xi_0)/z_j$; $\tau=m_j$, *the inequalities of* (3.8) *and/or* (3.9) *are fulfilled for* $t=0$ / $t=3$ / $t=0,1,2,3$.

Moreover we have the following theorem.

Theorem 5. *Let the data* (2.1) - (3.6) *be given. There are integers* $\bar{n}_{i,j}$, $\bar{m}_{i,j}$ *such that the interpolating polynomial* $b_{i,j} \in BP_{\bar{n}_{i,j},\bar{m}_{i,j}}$ *is shape-preserving. Let* $\bar{n}_{i,j}$, $\bar{m}_{i,j}$ *be any of such couple of integers. Then for any* $n_{i,j} \geq \bar{n}_{i,j}$, $m_{i,j} \geq \bar{m}_{i,j}$ *the interpolating* $b_{i,j} \in BP_{n_{i,j},m_{i,j}}$ *is shape preserving.*

It is clear that, if we construct an interpolating surface using the polynomials $b_{i,j} \in BP_{n_{i,j},m_{i,j}}$ for any rectangle $R_{i,j}$, this, in general, will not be either continous or differentiable across the grid lines, because of the different spaces adopted. On the other hand, if two adjacent rectangles $R_{i,j}$ and $R_{i+1,j}$ have the same degree in the y variable, then there will be a C^1 continuity across the common edge, since the two polynomials share the same interpolatory conditions. Thus, for obtaining a shape-preserving surface we construct the sequences $n_0,n_1,\dots,n_{N-1}$, $m_0,m_1,\dots,m_{M-1}$ using the following algorithm

1. *For* $j=0,1,\dots,M-1$
   1.1. *For* $i=0,1,\dots,N-1$
      1.1.1. *Check the kind of data in* $R_{i,j}$ *and compute the values* $n_{i,j}$, $m_{i,j}$ *(which allow shape-preserving interpolation) according to Theorem 4.*
2. *For* $i=0,1,\dots,N-1$
   2.1. Set $n_i=max\{n_{i,j}$; $j=0,1,\dots,M-1\}$.

3. For $j=0,1,\dots,M-1$

3.1 Set $m_j = max(m_{i,j}\ ;\ i=0,1,\dots,N-1)$

and construct the surface patching toghether the interpolating polynomials of BP_{n_i,m_j} that is taking an interpolating spline of

$$BS = \left\{ s\in C^1([x_0,x_N]x[y_0,y_M]) : s\in BP_{n_i,m_j} \text{ all } (x,y)\in R_{i,j}\ ;\ i=0,\dots,N-1\ ;\ j=0,\dots,M-1 \right\}.$$

We observe that, with the term *"shape-preservation"* we mean, infact, a *"selected"* shape-preservation or, in other words, that the polynomial has co-monotonicity and/or co-convexity properties *if these are requested by the user*. Suppose, for example, that one is interested in preserving the monotonicity in x and the convexity in y. Then the problems are a) to check the data in $R_{i,j}$ for verifying if the constraints of the required shape are fulfilled along some edge or inside the rectangle, and b) to verify if (or, as we will see later, to operate in such a way that) the polynomial is monotone in x and convex in y in the same regions of the data (as we will see later, the code SPBIS1 ([13]) is able to treat independently constraints on the shape of $b_{i,j}$ with respect to x and y; the possible options are no constraints, monotonicity constraints, convexity constraints or monotonicity and convexity constraints). In general, not all the data (2.1), (3.6) are available but only a subset, and, more frequently, only the points (3.6). It is possible to compute the partial derivatives of (3.6) *"line by line"* along the sections of equations $y=y_j$, $j=0,1,\dots,M$ and $x=x_i$, $i=0,1,\dots,N$ respectively, in such a way that their values agree with the shape of the corresponding sections of the points (2.1). Since in our methods the *"main variables"* are the degrees, a good choice of the derivatives is important but not crucial, and 1-D strategies can be used with a certain confidence. With more details we used the Bessel interpolation scheme (see, e.g. [9]) when no constraints are imposed on the shape, the local formulas of [12] for monotonicity constraints, and the sequential methods of [11] for convexity or monotonicity and convexity constraints. In these computations the degrees are set equal to certain values n, m if these are given in input, or equal to the minimal degree if they are automatically computed by the code. The second mixed derivatives of (3.6) are simply computed using a weighted average of the divided differences of the adjacent first partial derivatives, where the weights are proportional to the reciprocal of the interval lenghts. In closing, we note that the all the results are stated for strict inequalities, since otherwise it could be impossible to consider separately the shape constraints for each rectangle. So, in order to reproduce the behaviour of linear or planar subsets of the data, it is necessary to reprocess the computed derivatives on the same regions. This is done using a one-pass strategy which is based upon heuristic considerations, and takes into account the constraints in both the directions. It is important to remark that there is no teorethical guarentee that the interpolating surface preserves the I (D) and/or Cnv (Cnc) of the data inside the rectangles $R_{i,j}$; however this is what generally happens in practice, as the graphical examples reported in [14] and in [13] show. The method outlined above is the basis of the

FORTRAN 77 code SPBSI1. This is divided in two parts, SPBSC1 (Shape-Preserving Bivariate Spline Computation), which computes all the parameters - degrees and/or derivatives - needed for the construction of the surface, and SPBSE1 (Shape-Preserving Bivariate Spline Evaluation), which evaluates the spline at a prescribed set of points. The structure of the code is given in fig. 4 , while the user's options and the results obtainable are explained in fig. 5. We evidentiate the fact that the above discussion is stated for $C^1(R)$ spline, but the general formulation (both for the 1-D and the 2-D case, see [10], [11], [12], [14] and [13]) has been given in $C^k(R)$. From a practical point of view, however, this generalization is not so important as it could appear, because all the j-th partial derivatives, $j=2,\ldots,k$, vanish across the boundaries of $R_{i,j}$ .The graphical features of the method result clearly from figs. 5-7. The exact values of the data sets are given in [14]; for sake of brevity, they are reported here in a graphical form.

We conclude this section with some observations. Practical experiments have shown that, in some pathological examples, the convexity constraints require large degrees $(n_i$, $m_j)$ in $R_{i,j}$. Although these hold for the strips $[x_i,x_{i+1}]$, $[y_j,y_{j+1}]$, the dimension of the space is not affected and the roundoff errors are not magnified. The drawback is that the interpolant tends, in this case, to a bilinear shape, and its graph can be no more visually pleasing. On the other hand, nice surfaces are obtained for medium values, e.g. 5 - 8 , even if we have no theoretical guarentee that the inequalities are fulfilled. As a consequence, in [13] the following methods are suggested: a) to impose a medium value for the maximum degree allowed for the spline and to let the degrees be automatically computed, or b) to assign medium input values for the degrees of the spline. Another drawback is given by the low approximation properties; infact, although it is impossible to develop error estimates without clearly specify the relationship between the derivatives in (3.6) and the function to be approximated, it may be argued that the order of approximation (as the mesh size tends to zero) tends to be linear as the degrees increase (the function ℓ tends to be the bilinear interpolant and the Bernstein polynomial tends to approximate linearly ℓ). On the other hand, when compared with the existing techniques, this method has the advantage of a greater flexibility. Infact it can be applied to mixed Lagrange-Hermite interpolation problems and can handle any kind of constraints on the shape. From a graphical point of view, it seems to have both the advantages of the monotonicity preserving methods of sect. 2.2 , and the shape-preserving properties of the approach explained in sect. 2.1 . Researches are under study for developing techniques which are local and are not dependent of the order of data processing.

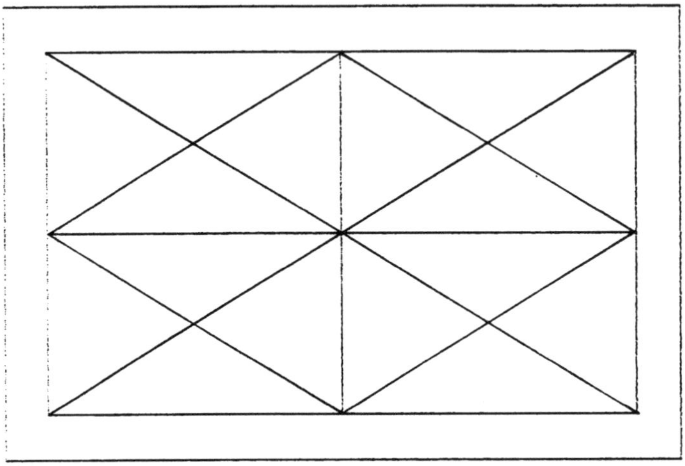

Fig. 1. The triangularization of $R_{i,j}$.

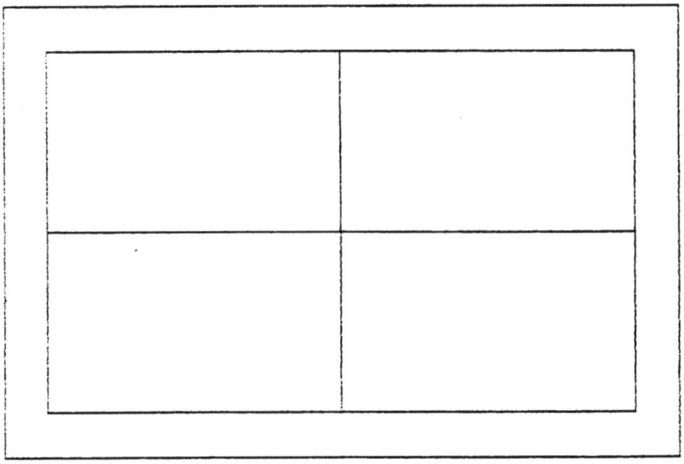

Fig. 2 The rectangular subdivision of $R_{i,j}$

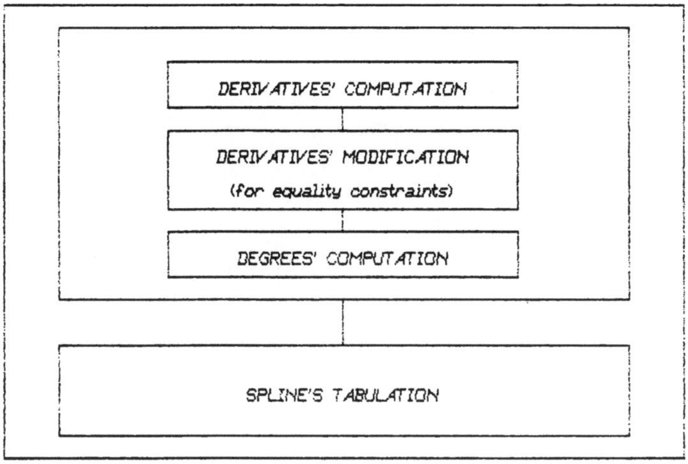

Fig. 3. The scheme of the code

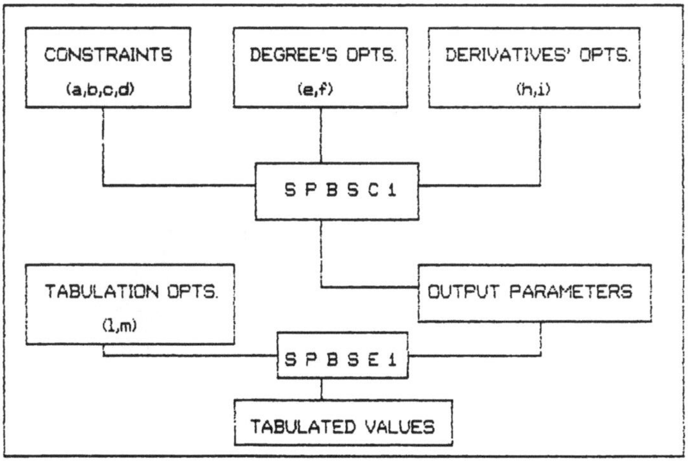

Fig. 4. The code features : a) no constraints; b) co-monotone; c) co-convex; d) co-monotone and co-convex; e/f) given in input/computed by the code; h/i) given in input/computed by the code (independently for the various derivatives); l/m) matrix of ordered points/array of irregularly spaced points.

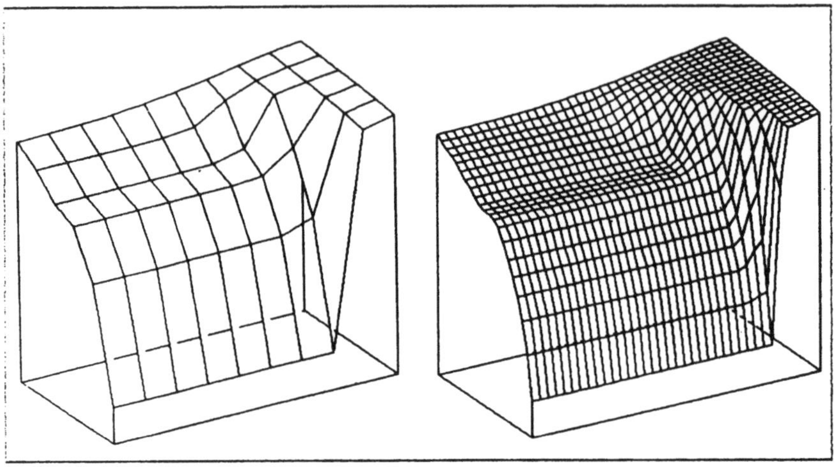

Fig. 5. The aluminum equation of state data (left) interpolated by a spline (right) co-onotone in both the variables

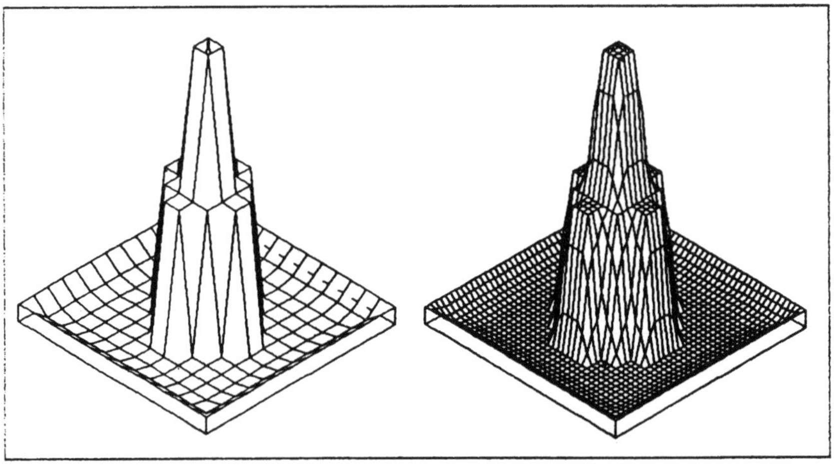

Fig. 6. The 'tower function' data (left) interpolated by a spline (right) co-monotone and co-convex in both the variables

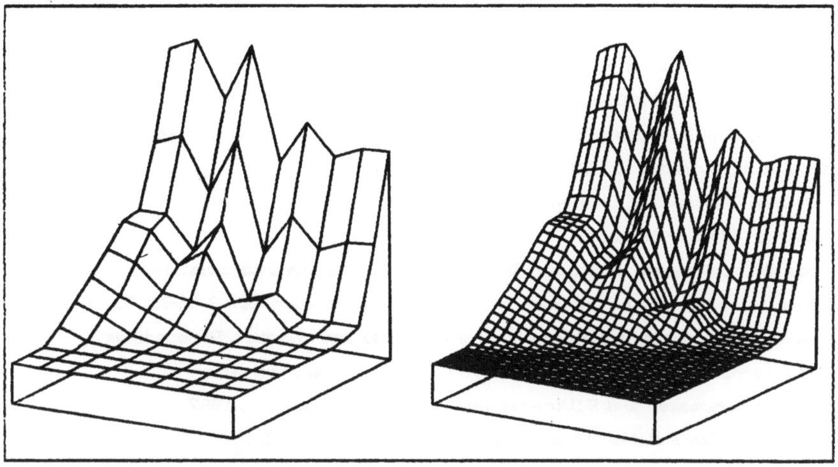

Fig. 7. The 'Akima test' data (left) interpolated by a spline (right) co-monotone and co-convex in x and co-monotone in y.

REFERENCES

[1] H.AKIMA, A method of bivariate interpolation and smooth surface fitting based on local procedures, *Comm. ACM* 17 (1974) 18-20.

[2] S.ASATURIAN, K.UNSWORTH, A C^1 monotonicity preserving surface interpolation scheme, University of Dundee, Dept. of Mathematics and Computer Sciences, Report CS 88/02.

[3] R.E.BARNHILL, *Blending function interpolation: a survey and some new results,* in Numerische Methoden der Approximationstheorie, Band 3, L.Collatz, H.Werner and G.Meinardus eds, ISNM vol.30, Birkhauser Verlag, Basel, 1976.

[4] R.K.BEATSON, Z.ZIEGLER, Monotonicity preserving surface interpolation, *SIAM J.Num.Anal.* 22 (1985) 401-411.

[5] W.BOHM, G.FARIN, J.KAHAMANN, A survey of curves and surface methods in CAGD, *CAGD* 1 (1984) 1-60.

[6] R.E.CARLSON, F.N.FRITSCH, Monotone piecewise bicubic interpolation, *SIAM J.Num.Anal.* 22 (1985) 386-410.

[7] R.E.CARLSON, F.N.FRITSCH, An algorithm for monotone piecewise bicubic interpolation, Lawrence Livermore Laboratory, Report UCRI-96791, 1987.

[8] C.K.CHUI, *Multivariate splines. Theory and applications,* CBMS - NSF Lectures, Texas A&M University, Department of Mathematics, 1987.

[9] S.D.Conte, C. de Boor, *Elementary numerical analysis. An algorithmical approach.* Mc Graw Hill Book Company, Singapore, 1981.

[10] P.Costantini, On monotone and convex spline interpolation, *Math.Comp.* 46(1986) 203-214.

[11] P.Costantini, An algorithm for computing shape-preserving interpolating splines of arbitrary degree, *J.Comp.Appl.Math* 22 (1988) 89-136.

[12] P.Costantini, Co-monotone interpolating splines of arbitrary degree. A local approach, *SIAM J.Sci.Stat.Comput.* 8 (1987) 1026-1034.

[13] P.Costantini, SPBSI1: a code for computing shape-preserving bivariate spline interpolation, *submitted to ACM Trans. on Math. Softw.*

[14] P.Costantini, F.Fontanella, Shape-preserving bivariate interpolation, *to appear on SIAM J.Num.Anal.*

[15] S.L.Dodd, D.F.Mc Allister, J.A.Roulier, Shape-preserving spline interpolation for specifying bivariate functions on grids, *IEEE Comp.Graph.Appl* 3 (1983) 70-79.

[16] R.F.Franke,L.L.Schumaker, *A bibliography of multivariate approximation*, Proocedings of the international workshop on applied multivariate approximation, C.K.Chui, L.L.Schumaker and F.I.Utreras eds., Academic Press, N.Y., 1987.

[17] F.N.Fritsch, R.E.Carlson, Monotone piecewise cubic interpolation, *SIAM J.Num.Anal.* 17 (1980) 238-246.

[18] F.N.Fritsch, R.E.Carlson, Monotonicity preserving bicubic interpolation. A progress report", *CAGD* 2 (1985) 117-121.

[19] W.J.Gordon, An operator calculus for surface and volume modeling, *IEEE Comp.Graph.Appl* 3 (1983) 18-22.

[20] G.G.Lorentz, *Bernstein polynomials*, University of Toronto Press, Toronto, 1953.

[21] D.F.Mc Allister, J.A.Roulier, Interpolation by convex quadratic splines, *Math.Comp.* 31 (1978) 1154-1162.

[22] D.F.Mc Allister, J.A.Roulier, An algorithm for computing a shape-preserving osculatory quadratic spline, *ACM Trans. Math. Softw.* 7 (1981) 331-347.

[23] C.A.Micchelli, F.I.Utreras, Smoothing and interpolation in a convex subset of a Hilbert space, *SIAM J.Sci.Stat.Comput* 9 (1988) 728-746.

[24] J.W.Shmidt, 1-D shape-preserving spline interpolation: existence theorems and construction of optimal splines, Technische Universitat Dresden, Sektion Mathematik, report 07-01-87, 1987.

[25] F.I.Utreras, Positive thin plate splines, *J.Aprox.Th. and Appl* 1 (1985) 77-108.

[26] F.I.Utreras, *Constrained surface construction*, Proocedings of the international workshop on applied multivariate approximation, C.K.Chui, L.L.Schumaker and F.I.Utreras eds., Academic Press, N.Y., 1987.

[27] F.I.Utreras, M.L.Varas, Monotone interpolation of scattered data in $\mathbb{R}^2$, Tech. Rep., Depto. Mat. Ap., Universidad de Chile, 1987.

Paolo Costantini, Dipartimento di Matematica, Via del Capitano, 15 - 53100 Siena, Italia

ON SPLINE APPROXIMATION FOR CONVOLUTION EQUATIONS

Johannes Elschner [1)]

1. Introduction

We consider the numerical solution of the Wiener-Hopf integral equation

$$Au(x) := u(x) - \int_0^\infty \varkappa(x-y)u(y)dy = f(x), \quad x \in \mathbb{R}^+ = [0,\infty) \qquad (1)$$

and the Mellin convolution equation

$$Bu(x) := u(x) - \int_0^1 k(x/y)y^{-1}u(y)dy = f(x), \quad x \in [0,1] \ , \qquad (2)$$

where f, $\varkappa$ and k are given functions and u is the unknown function. Such equations arise in a number of applications; for example, (1) occurs in radiative equilibrium and transfer, and (2) appears when boundary integral methods are used in potential theory or elastostatics (cf. [3], [5] and the references there).

In this paper we give a survey of recent results on stability and convergence of Galerkin and collocation methods based on piecewise polynomials for Eqs. (1) and (2). We also consider the corresponding iterated Galerkin solutions as well as the discrete and iterated versions of the collocation methods. Note that the integral operators

$$Wu(x) := \int_0^\infty \varkappa(x-y)u(y)dy \ , \quad Ku(x) := \int_0^1 k(x/y)y^{-1}u(y)dy$$

in (1) and (2) are not compact so that standard theories for the numerical analysis of second kind integral equations (cf. [1]) cannot be applied. Nevertheless, using graded meshes and modified spline spaces, one obtains the same optimal orders of convergence as in the case of second kind equations with smooth or weakly singular kernels on finite intervals (cf. e.g. [1], [13]).

For any interval $J \subset \mathbb{R}$ and $1 \le p \le \infty$, $L_p(J)$ will denote the usual Lebesgue space on J with norm

$$\|u\|_{p,J} = \{ \int_J |u|^p dx \}^{1/p}, \ p<\infty; \quad \|u\|_{\infty,J} = \text{ess sup}_J |u(x)| \ .$$

For $l \in \mathbb{N}$ and $\varrho \ge 0$, we further introduce the weighted Sobolev spaces

$$W_p^{l,\varrho}(J) = \{u \in \mathcal{D}'(J): e^{\varrho x}D^j u \in L_p(J) \ , \ j=0,\dots,l\} \ ,$$

$$L_p^{l,\varrho}(J) = \{u \in \mathcal{D}'(J): |x|^{j-\varrho} D^j u \in L_p(J) \ , \ j=0,\dots,l\}$$

1) Karl-Weierstraß-Institut für Mathematik, Akademie der Wissenschaften der DDR, Mohrenstr. 39, Berlin, DDR-1086

equipped with the canonical norms, where $D = d/dx$ and the ϱ is dropped from the notation if $\varrho = 0$.

With the Wiener-Hopf integral operator A defined by (1') we associate its symbol $\alpha(\xi) = 1 - a(\xi)$, where a is the Fourier transform of the kernel $\varkappa$:

$$a(\xi) = \int_{-\infty}^{\infty} e^{ix\xi} \varkappa(x)\, dx .$$

We now introduce the following assumptions on $\varkappa$ and α.

(V1): $\varkappa \in W_1^l(\mathbb{R})$ for all $l \in \mathbb{N}$;

$$\alpha(\xi) \neq 0,\ \xi \in \mathbb{R} ;\ \{\arg \alpha(\xi)\}_{-\infty}^{\infty} = 0 .$$

(V2): There is some $\varrho > 0$ such that $\varkappa \in W_1^{l,\varrho}(\mathbb{R})$ for all $l \in \mathbb{N}$ and

$$\alpha(\xi) \neq 0 \quad \text{for } -\varrho \leq \operatorname{Im} \xi < 0 .$$

Here $\{\arg \cdot\}_{-\infty}^{\infty}$ denotes the variation of the argument when ξ runs from $-\infty$ to ∞. (V1) guarantees that A is continuous and invertible on each of the spaces $L_p(\mathbb{R}^+)$, $1 \leq p \leq \infty$, and W is a continuous map of $L_p(\mathbb{R}^+)$ into $W_p^l(\mathbb{R}^+)$ for any l, whereas (V2) ensures boundedness and invertibility of A in $W_p^{l,\varrho}(\mathbb{R}^+)$, $l \in \mathbb{N}$ (see [9] and [7]).

Let b be the Mellin transform of the kernel k of K :

$$b(z) = \int_0^{\infty} x^{z-1} k(x)\, dx .$$

Then the function $\beta(z) = 1 - b(z)$ is called the symbol of the Mellin convolution operator B defined by (2). For fixed p, $1 \leq p \leq \infty$, we make the following set of assumptions on k and β. Let $p' = p/(p-1)$.

(A1): $k \in L_1^{l,1/p'}(\mathbb{R}^+)$ for all $l \in \mathbb{N}$;

$$\beta(z) \neq 0,\ \operatorname{Re} z = 1/p ;\ \{\arg \beta(1/p + i\xi)\}_{-\infty}^{\infty} = 0 .$$

(A2): There exists some $\varrho > 0$ such that $k \in L_1^{l,1/p'+\varrho}(\mathbb{R}^+)$ for all $l \in \mathbb{N}$ and $\beta(z) \neq 0$ for $1/p - \varrho \leq \operatorname{Re} z < 1/p$.

(A1) implies that B is a continuous and invertible operator on $L_p(0,1)$ and K is a bounded map of $L_p(0,1)$ into $L_p^l(0,1)$ for any l, whereas (A2) ensures continuity and invertibility of B on the spaces $L_p^{l,\varrho}(0,1)$, $l \in \mathbb{N}$ (cf. [7]). Note that the supremum of the numbers $\varrho > 0$ for which condition (A2) is satisfied reflects the principal term of the asymptotics of solutions to Eq. (2) at the point x=0 (cp. [6]).

Example. When the double layer potential is used to solve the Dirichlet problem and the (plane) domain contains a corner with an interior angle $(1-\lambda)\pi$, $\lambda \in (-1,1)$, an integral equation (2) with the kernel function

$$k(x) = \pi^{-1} x \sin\lambda\pi / (1 + 2x\cos\lambda\pi + x^2)$$

appears (cf.[5]). Then the symbol of B takes the form $1 + \sin\lambda\pi z / \sin\pi z$, and it can be checked that condition (A1) is valid for $1<p\leq\infty$ and (A2) holds for $\varrho < 1/p + 1/(1+|\lambda|)$.

Finally, we remark that Eqs. (1) and (2) are related by a simple transformation of the interval [0,1] to the half-axis. Indeed, if we consider the operator B (on L_p) with kernel k and symbol β and take the isomorphism $Ru(x) = e^{-x/p}u(e^{-x})$ of $L_p(0,1)$ onto $L_p(\mathbb{R}^+)$, we observe that RBR^{-1} is a Wiener-Hopf operator with kernel $\varkappa(x) = e^{-x/p}k(e^{-x})$ and symbol $\alpha(\xi) = \beta(1/p + i\xi)$.

2. Wiener-Hopf equations

Let Δ^n be a sequence of meshes $\{x_i^{(n)}: 0\leq i\leq n\}$ on $\mathbb{R}^+$, where $0=x_0^{(n)}< x_1^{(n)}< \ldots < x_n^{(n)} = \infty$. For convenience write x_i for $x_i^{(n)}$ where possible. Let $q\geq 1$. We shall say that Δ^n is of class $\mathcal{M}_q$ ($\Delta^n\in\mathcal{M}_q$) if

$$c^{-1}(n-i)^{q-1}/n^q \leq \exp(-x_i^{(n)}) - \exp(-x_{i+1}^{(n)}) \leq c(n-i)^{q-1}/n^q, \quad i=0,\ldots,n-1$$

with some constant $c>0$ independent of i and n. A typical example is given by $x_i^{(n)} = q\log(n/(n-i))$. For a sequence of partitions Δ^n, we further introduce the modified spline spaces

$$S_d(\Delta^n) = \{u\in L_\infty(\mathbb{R}^+): u|_{(x_{i-1},x_i)}\in \mathbb{P}_d\ ,\ i=1,\ldots,n-i_0\ ,\ u|_{(x_{n-i_0},\infty)} = 0\},$$

where the index $i_0\geq 1$ does not depend on n and $\mathbb{P}_d$ denotes the set of polynomials of degree $\leq d$.

For the approximate solution of Eq. (1), we first consider the classical Galerkin method with splines from $S_d(\Delta^n)$. The Galerkin solution $u_n\in S_d(\Delta^n)$ is defined by

$$P^nAu_n = u_n - P^nWu_n = P^nf\ , \tag{3}$$

where P^n is the orthogonal projection of $L_2(\mathbb{R}^+)$ onto $S_d(\Delta^n)$.

<u>Theorem 1.</u> Assume $1\leq p\leq\infty$, (V1), $\Delta^n\in\mathcal{M}_q$, and suppose that the index i_0 is sufficiently large. Then the Galerkin method (3) is stable in $L_p(\mathbb{R}^+)$, i.e.

$$\|P^nAu_n\|_{p,\mathbb{R}^+} \geq c\|u_n\|_{p,\mathbb{R}^+}\ ,\ u_n\in S_d(\Delta^n)\ ,\ n\geq n_0 \tag{4}$$

when n_0 is large enough.

This result was proved in [7], using the Wiener-Hopf factorisation of the operator A and local approximation properties of the spline spaces. We remark that the theorem does not hold for $i_\sigma=1$, in general. Estimate (4) implies that the Galerkin equations (3) are uniquely solvable for any $f\in L_p(\mathbb{R}^+)$ if $n\geq n_0$. The next result shows that one obtains the optimal order of convergence when suitably graded meshes are used. Let u be the exact solution of Eq. (1).

Theorem 2. [7] Assume $1\leq p\leq\infty$, (V1), (V2), $\Delta^n\in\mathcal{M}_q$, $q\geq(d+1)/\varrho$, $f\in W_p^{d+1,\varrho}(\mathbb{R}^+)$, and suppose that i_0 is large enough. Then the method (3) converges with the error bound

$$\|u - u_n\|_{p,\mathbb{R}^+} = O(n^{-d-1}) \quad \text{as} \quad n\to\infty \; . \tag{5}$$

The iterated Galerkin solution is now defined by $u_n^* = f + Wu_n$, where u_n is given by (3). We have the following superconvergence result.

Theorem 3. [7] If the assumptions of the preceding theorem are satisfied and the stronger grading $q\geq 2(d+1)/\varrho$ is given, then $\|u - u_n^*\|_{p,\mathbb{R}^+} = O(n^{-2d-2})$ as $n\to\infty$.

Next we consider collocation methods based on piecewise polynomials. Set $x_{ij} = x_{i-1}+\xi_j(x_i-x_{i-1})$, where ξ_j are fixed points with $0\leq\xi_0<\xi_1<\ldots<\xi_d\leq 1$, and for any continuous function u on $\mathbb{R}^+$, define the interpolatory projection $Q^n u\in S_d(\Delta^n)$ by

$$(Q^n u)(x_{ij}) = u(x_{ij}),\; i=1,\ldots,n-i_0,\; j=0,\ldots,d \; .$$

We seek an element $u_n\in S_d(\Delta^n)$ such that

$$Q^n Au_n = u_n - Q^n Wu_n = Q^n f \; . \tag{6}$$

Theorem 4. Under the assumptions of Theorem 1, the collocation method (6) is stable in $L_p(\mathbb{R}^+)$, i.e. estimate (4) holds with P^n replaced by Q^n. Moreover, if the conditions of Theorem 2 are satisfied, then (6) converges in $L_p(\mathbb{R}^+)$ with the error bound (5).

We now discuss the iterated collocation (or product integration) method. The iterated collocation solution is defined by $u_n^* = f + Wu_n$, where u_n is given by (6). Its convergence rate depends on the choice of the points ξ_j. Introduce the (d+1)-point interpolatory quadrature rule on [0,1]

$$\int_0^1 v\,dx \approx \sum_{j=0}^{d} \omega_j v(\xi_j) \tag{7}$$

with weights ω_j and assume that $R = d+1+r$ is the order of this rule, i.e. it is exact for all polynomials of degree $< R$ on [0,1]. Note that $0\leq r\leq d+1$ and $r = d+1$ if and only if ξ_j are the Gauss-Legendre points on [0,1]. We have the following superconvergence result.

Theorem 5. Assume $1\le p\le\infty$, (V1), (V2), $\Delta^n\in\mathcal{M}_q$, $q\ge R/\varrho$, $f\in W_p^{R,\varrho}(\mathbb{R}^+)$, and suppose that i_0 is large enough. Then $\|u-u_n^*\|_{p,\mathbb{R}^+}=O(n^{-R})$, $n\to\infty$.

In the case $p=\infty$ the proof of Theorems 4 and 5 can be found in [7]; for the general case, see [8]. Some results on collocation with piecewise constant basis functions were obtained by Sloan and Spence [12].

Finally, we consider certain quadrature methods based on piecewise polynomials. The composite quadrature rule obtained by shifting (7) to each subinterval (x_{i-1},x_i) of Δ^n, and summing over $i\le n-i_0$, is

$$\int_0^\infty u\,dx \approx \sum_{(i,j)\in\mathcal{J}} \omega_j u(x_{ij})(x_i-x_{i-1}),$$

where $\mathcal{J}:=\{(i,j):\ 1\le i\le n-i_0,\ 0\le j\le d\}$. Thus we approximate the integral operator W in (1) by

$$W_n u(x) = \sum_{(i,j)\in\mathcal{J}} \omega_j \varkappa(x-x_{ij})u(x_{ij})(x_i-x_{i-1}) .$$

For the integral equation (1), the discrete collocation method is the following: find $u_n\in S_d(\Delta^n)$ such that

$$u_n - Q^nW_nu_n = Q^nf \quad . \tag{8}$$

The discrete iterated (or Nyström) solution u_n^* to (1) is then defined by

$$u_n^* = f + W_nu_n \quad . \tag{9}$$

Note that $Q^nu_n^*=u_n$. (8) is a linear system in the values $u_n^*(x_{ij})$, and then the right-hand side of (9) may be computed giving $u_n^*(x)$ for all x.

Theorem 6. [8] Under the conditions of Theorem 1, the discrete collocation method (8) is stable in $L_p(\mathbb{R}^+)$, i.e. inequality (4) holds with P^nA replaced by $Q^n(I-W_n)$. Furthermore, if the assumptions of Theorem 2 are satisfied, then (8) converges with the error estimate (5).

Theorem 7. [8] Under the assumptions of Theorem 5, the Nyström method (9) converges with the error bound $\|u-u_n^*\|_{p,\mathbb{R}^+}=O(n^{-R})$, $n\to\infty$.

In the case $p=\infty$, Theorem 7 is due to Chandler and Graham [3].

3. Mellin convolution equations

Let $\Delta_n=\{x_i:\ 0\le i\le n\}$, $x_i=x_i^{(n)}$, be a sequence of meshes, where $0=x_0^{(n)} < x_1^{(n)} < \dots < x_n^{(n)}=1$. By definition, the sequence Δ_n is of class $\mathcal{M}_q$, $q\ge 1$, if

$$c^{-1}i^{q-1}/n^q \le x_i^{(n)} - x_{i-1}^{(n)} \le c\, i^{q-1}/n^q \quad , \quad i=1,\dots,n$$

for some constant c which does not depend on i and n. A typical example is given by the graded meshes

$$x_i^{(n)} = (i/n)^q \quad , \ i=0,\dots,n \ , \quad n\in\mathbb{N} \ . \tag{10}$$

To a sequence of partitions Δ_n we further associate the modified spline spaces

$$S_d(\Delta_n) = \{u \in L_\infty(0,1): u|_{(0,x_{i_0})} = 0,\ u|_{(x_{i-1},x_i)} \in \mathbb{P}_d,\ i=i_0+1,\dots,n\},$$

where the index $i_0 \geq 1$ is independent of n. Let P_n be the orthogonal projection of $L_2(0,1)$ onto $S_d(\Delta_n)$. The Galerkin method based on piecewise polynomials for the approximate solution of Eq. (2) is the following: find $u_n \in S_d(\Delta_n)$ such that

$$P_nBu_n = u_n - P_nKu_n = P_nf \ . \tag{11}$$

Let u be the exact solution of Eq. (2).

Theorem 8. [7] (i) Assume (A1), $\Delta_n \in \mathcal{M}_q$, and suppose that the index i_0 is sufficiently large. Then the Galerkin method (11) is stable in $L_p(0,1)$, i.e.

$$\| P_nBu_n \|_{p,(0,1)} \geq c \|u_n\|_{p,(0,1)} \quad , u_n \in S_d(\Delta_n) \ , n \geq n_0 \tag{12}$$

when n_0 is large enough.
(ii) If, in addition, (A2), $q \geq (d+1)/\rho$ and $f \in L_p^{d+1,\rho}(0,1)$ hold, then (11) converges with the error bound

$$\| u - u_n \|_{p,(0,1)} = O(n^{-d-1}) \quad \text{as} \quad n \to \infty \ . \tag{13}$$

Moreover, for the iterated Galerkin solution $u_n^* = f + Ku_n$, we have the following superconvergence result.

Theorem 9. [7] If the assumptions (A1), (A2), $\Delta_n \in \mathcal{M}_q$, $q \geq 2(d+1)/\rho$ and $f \in L_p^{d+1,\rho}(0,1)$ are satisfied and the index i_0 is large enough, then

$$\|u - u_n^*\|_{p,(0,1)} = O(n^{-2d-2}) \quad \text{as} \quad n \to \infty \ .$$

For p=2 and $p=\infty$, Theorems 8 and 9 were obtained by Chandler and Graham [2],[5] under the additional assumption that the corresponding operator norms of K are smaller than one. In the latter case, however, the stability of (11) already holds with $i_0=1$ (i.e. without modification of the spline spaces), whereas this is not true in the general case (cf. [7, Sec. 3]).

Define the collocation points $x_{ij} = x_{i-1} + \xi_j(x_i - x_{i-1})$ as in Sec. 2, and consider the interpolatory projections $Q_nu \in S_d(\Delta_n)$ given by

$$(Q_nu)(x_{ij}) = u(x_{ij}),\ (i,j) \in \mathcal{J}\ ,\ \mathcal{J} := \{(i,j): i_0 < i \leq n,\ 0 \leq j \leq d\}\ ,$$

where u is an arbitrary continuous function on (0,1]. Our collocation method for the solution of Eq. (2) is the following: find $u_n \in S_d(\Delta_n)$ such that

$$Q_nBu_n = u_n - Q_nKu_n = Q_nf \ . \tag{14}$$

The corresponding iterated collocation (or product integration) solution is then defined by $u_n^* = f + Ku_n$. The proof of the following two assertions can be found in [8]; they extend results from [5] and [7].

<u>Theorem 10.</u> If the conditions of Theorem 8(i) are satisfied, then the collocation method (14) is stable in $L_p(0,1)$, i.e. estimate (12) holds with P_n replaced by Q_n. Furthermore, under the assumptions of Theorem 8(ii), (14) converges in $L_p(0,1)$ with the error estimate (13).

<u>Theorem 11.</u> Let R be the order of the quadrature rule (7). Assume (A1), (A2), $\Delta_n \in \mathcal{M}_q$, $q \geq R/\varrho$, $f \in L_p^{R,\varrho}(0,1)$, and suppose that i_0 is large enough. Then $\| u - u_n^* \|_{p,(0,1)} = O(n^{-R})$ as $n \to \infty$.

Note that the maximal order $O(n^{-2d-2})$ in the last theorem is achieved by selecting the Gauss-Legendre points ξ_j on $[0,1]$ for the definition of the collocation points.

To conclude this section, we discuss discrete collocation and discrete iterated collocation (or Nyström) methods based on piecewise polynomials. With the rule (7) we associate the composite rule

$$\int_0^1 u \, dx \approx \sum_{(i,j)\in \mathcal{J}} \omega_j u(x_{ij})(x_i - x_{i-1}) .$$

Using this we approximate the integral operator K in (2) by

$$K_n u(x) = \sum_{(i,j)\in \mathcal{J}} \omega_j k(x/x_{ij}) x_{ij}^{-1} (x_i - x_{i-1}) u(x_{ij}) .$$

The discrete collocation solution $u_n \in S_d(\Delta_n)$ to (2) is then defined by

$$u_n - Q_n K_n u_n = Q_n f \quad , \tag{15}$$

and the Nyström solution to (2) is given by

$$u_n^* = f + K_n u_n \; . \tag{16}$$

<u>Theorem 12.</u>[8] Under the conditions of Theorem 8(i), the iterated collocation method (15) is stable in $L_p(0,1)$, i.e. estimate (12) holds with $P_n B$ replaced by $Q_n(I-K_n)$. Moreover, if the assumptions of Theorem 8(ii) are satisfied, then (15) converges with the error bound (13).

<u>Theorem 13.</u>[8] Under the assumptions of Theorem 11, the Nyström method (16) converges with the error estimate $\| u - u_n^* \|_{p,(0,1)} = O(n^{-R}), n \to \infty$

The last theorem generalizes a result proved in [4] in the case $p = \infty$. Finally, we remark that Rathsfeld [11] and Prößdorf and Rathsfeld [10] investigated certain spline collocation and quadrature methods for a more general class of Mellin convolution operators, which contains integral equations of the form (2) and singular integral operators of Cauchy type. Using Mellin transformation techniques and local principles for projection methods, they obtained sharp stability results in the case of the special meshes (10).

References

[1] K.E.Atkinson, A survey of numerical methods for the solution of Fredholm integral equations of the second kind, SIAM, Philadelphia 1976 .

[2] G.A.Chandler, I.G.Graham, Uniform convergence of Galerkin solutions to non-compact integral operator equations, IMA J. Numer. Anal. 7 (1987) 327-334 .

[3] ---,---, The convergence of Nyström methods for Wiener-Hopf equations, Numer. Math. 52 (1988) 345-364

[4] ---,---, Higher order methods for linear functionals of solutions of second kind integral equations, SIAM J.Numer. Anal. (to appear)

[5] ---,---, Product integration-collocation methods for non-compact integral operator equations, Math. Comp. 50 (1988) 125-138 .

[6] J.Elschner, Asymptotics of solutions to pseudodifferential equations of Mellin type, Math. Nachr. 130 (1987) 267-305 .

[7] ---, On spline approximation for a class of non-compact integral equations, Math. Nachr. (to appear) .

[8] ---, On spline collocation for convolution equations (in preparation) .

[9] I.Z.Gochberg, I.A.Feldman, Faltungsgleichungen und Projektionsverfahren zu ihrer Lösung, Akademie-Verlag, Berlin 1974 .

[10] S.Prößdorf, A.Rathsfeld, Mellin techniques in the numerical analysis of one-dimensional singular integral equations, Preprint P-Math-06/88, Karl-Weierstraß-Inst. Math.Akad.Wiss. DDR, Berlin 1988 .

[11] A.Rathsfeld, Eine Quadraturformelmethode für Mellin-Operatoren nullter Ordnung, Math. Nachr. 137 (1988) .

[12] I.H.Sloan, A.Spence, Projection methods for integral equations on the half-line, IMA J. Numer.Anal. 6 (1986) 153-172 .

[13] G.Vainikko, A.Pedas, P.Uba, Methods for solving weakly singular integral equations (Russian), Gos. Univ. Tartu . 1984 .

BIVARIATE INTERPOLATING AND SMOOTHING TENSOR PRODUCT SPLINES

S.Ewald[1], H.Mühlig[2], B.Mulansky[2]

1. Introduction

A well known and useful principle for the transfer of univariate linear approximation methods to bivariate problems consists in the formation of their tensor product schemes. These schemes are restricted to data given on a rectangular grid. On the other hand, they possess the main advantage, that the computations separate into the solution of a number of univariate approximation problems. For a complete discussion of the tensor product scheme of general interpolation methods and for some explanations of its limits and advantages we refer to [3, Chapter XVII].

The popularity of interpolating splines mainly stems from their extremal properties. For instance, the classic result of Holladay shows, that the natural cubic interpolating spline appears as the unique solution of the problem

$$\text{minimize} \int_a^b [D^2x(t)]^2dt \quad \text{over} \quad x \in \{x \in W_2^2[a,b]: x(\tau_i)=z_i \quad (i=1,\dots,m)\}.$$

This extremal property inspired the variational approach to splines in Hilbert spaces, see [2]. The abstract approach also includes multivariate splines. But we are especially interested in the identification of the extremal properties, which are related to the constructive tensor product of two spline interpolation operators. As a first attempt in this direction, it was incorrectly stated in [1], that the natural bicubic interpolating spline is the unique solution of the problem

$$\text{minimize} \int_a^b \int_{\tilde{a}}^{\tilde{b}} [D^{2,2}x^o(t,\tilde{t})]^2 d\tilde{t}\,dt$$

$$\text{over} \quad x^o \in \{x^o \in W_2^{2,2}([a,b]\times[\tilde{a},\tilde{b}]): \quad x^o(\tau_i,\tilde{\tau}_j)=z^{\bullet}_{ij} \quad (i=1,\dots,m, \ j=1,\dots,\tilde{m})\}.$$

[1] Kombinat VEB Carl Zeiss Jena, Betrieb Entwicklung wissenschaftlich-technischer Ausrüstungen, Göschwitzer Str. 33, Jena-Göschwitz, 6905, DDR

[2] Technische Universität Dresden, Sektion Mathematik, Wissenschaftsbereich Numerische Mathematik, Mommsenstr.13, Dresden, 8027, DDR

In [8], [16], [7], [17] this result was corrected and generalized to bivariate polynomial and L-splines with different boundary conditions. In section 3 of this paper we study the tensor product scheme of abstract interpolating splines in Hilbert spaces. To this end we give a short introduction to the topological tensor product of Hilbert spaces and of continuous linear operators in section 2. We restrict to the case of real function spaces in order to be able to give a more descriptive definition of tensor products. The introduced notions are illustrated by some examples related to bicubic splines.
The main result consists in an extremal property, which uniquely characterizes the constructive tensor product of two spline interpolation operators. We point out, that some modifications of the objective functional are possible without affecting the solution. This result contains all the investigations mentioned above as special cases.

In many applications, the strong interpolation conditions are not adequate, since the given data are affected by errors. Therefore, the notion of smoothing splines was introduced. For instance, the natural cubic smoothing spline is defined as the solution of the problem

$$\text{minimize} \int_a^b [D^2x(t)]^2 dt + \rho \cdot \sum_{i=1}^{m} \omega_i [x(\tau_i)-z_i]^2 \quad \text{over} \quad x \in W_2^2[a,b],$$

see, e.g., [12], [13].
Much effort has been done to find appropriate objective functionals for bivariate smoothing splines in order to reach the advantage of a separating solution. On the basis of some results for special cases in [15], [16], [4], [6], [7], where the two papers mentioned at last especially stimulated our investigations, in section 4 we consider the full tensor product scheme of two abstract spline smoothing operators. It is shown, that the computation of the corresponding smoothing spline separates in case of the introduced objective functional. Again, our result contains the mentioned investigations as particular cases.

The considerations presented in this paper give a homogeneous deduction of the known results for special types of interpolating and smoothing tensor product splines. Therefore, our results reveal the underlying principles, on which the separation of the solution of the corresponding extremal problem is based. They enable the creation of new tensor product schemes by the combination of arbitrary types of interpolating and smoothing univariate splines.
In this paper we sketch the investigations of the important examples of natural and complete bicubic interpolating and smoothing splines in order to illustrate the abstract approach.

2. Preliminaries

Tensor products. We need the notion of the (topological) tensor product of Hilbert spaces. Therefore we give now a short discussion of the tensor product of Hilbert spaces of real-valued functions, since in this special case it is possible to introduce the basic ideas in a more descriptive manner, compare [3, Chapter XVII]. For further informations about the topological tensor product of Hilbert spaces we refer the reader to [10], [11] and [14].

Let X and $\tilde{X}$ be Hilbert spaces of real-valued functions defined on T and $\tilde{T}$, respectively. The tensor product $x \otimes \tilde{x}$ of two functions $x \in X$, $\tilde{x} \in \tilde{X}$ is a function on $T \times \tilde{T}$ defined by the rule

$$x \otimes \tilde{x}\,(t,\tilde{t}) := x(t)\cdot\tilde{x}(\tilde{t}) \qquad \text{for } (t,\tilde{t}) \in T \times \tilde{T}. \tag{2.1}$$

The algebraic tensor product $X \otimes \tilde{X}$ of the linear function spaces X and $\tilde{X}$ is obtained as the linear span of all functions of the form (2.1), that means

$$X \otimes \tilde{X} := \{\sum_{i=1}^{n} \alpha_i\cdot(x_i \otimes \tilde{x}_i) \;:\; x_i \in X,\ \tilde{x}_i \in \tilde{X},\ n \in N\}. \tag{2.2}$$

With respect to this construction the space $X \otimes \tilde{X}$ is endowed in a natural way with an inner product, which is obtained from

$$(x^1\otimes\tilde{x}^1, x^2\otimes\tilde{x}^2)_{X\otimes\tilde{X}} := (x^1,x^2)_X(\tilde{x}^1,\tilde{x}^2)_{\tilde{X}} \qquad \text{for } x^1,x^2 \in X \text{ and } \tilde{x}^1,\tilde{x}^2 \in \tilde{X} \tag{2.3}$$

by bilinear extension.
Throughout the paper the notation $(.,.)_X$ denotes the inner product in a space X and $\|.\|_X$ the corresponding norm.
The topological tensor product $X \hat{\otimes} \tilde{X}$ of X and $\tilde{X}$ is defined as the completion of $X \otimes \tilde{X}$ with respect to the topology induced by the inner product defined above.

Let X, $\tilde{X}$, Z, $\tilde{Z}$ be Hilbert spaces of real functions and $A: X \to Z$, $\tilde{A}: \tilde{X} \to \tilde{Z}$ continuous linear operators. Then it is natural to define the tensor product $A \hat{\otimes} \tilde{A} : X \hat{\otimes} \tilde{X} \to Z \hat{\otimes} \tilde{Z}$ of A and $\tilde{A}$ by linear and continuous extension using the basic rule

$$A \hat{\otimes} \tilde{A}\,(x\otimes\tilde{x}) := Ax \otimes \tilde{A}\tilde{x} \qquad \text{for } x \in X,\ \tilde{x} \in \tilde{X}. \tag{2.4}$$

For the justification of the introduced constructions we refer especially to [3] and [11].

In order to be well prepared for our following considerations we also sketch the tensor product scheme of product spaces. To this end let $W := Y \times Z$ be endowed with the inner product

$$(w^1,w^2)_W := (y^1,y^2)_Y + \rho\cdot(z^1,z^2)_Z, \tag{2.5}$$

where $\rho > 0$ and $w^i = (y^i,z^i) \in W = Y \times Z$, $i=1,2$.
Let the space $\tilde{W}$ and its inner product be defined analogously.
We get

$$W \hat{\otimes} \tilde{W} = (Y \hat{\otimes} \tilde{Y}) \times (Y \hat{\otimes} \tilde{Z}) \times (Z \hat{\otimes} \tilde{Y}) \times (Z \hat{\otimes} \tilde{Z}). \tag{2.6}$$

By this result the notation

$$w = \begin{pmatrix} w_{11} & w_{12} \\ w_{21} & w_{22} \end{pmatrix}, \text{ where } w_{11} \in Y\hat{\otimes}\tilde{Y},\quad w_{12} \in Y\hat{\otimes}\tilde{Z},\quad w_{21} \in Z\hat{\otimes}\tilde{Y}$$

and $w_{22} \in Z\hat{\otimes}\tilde{Z}$, is inspired for $w \in W \hat{\otimes} \tilde{W}$.

Now, the inner product in $W \hat{\otimes} \tilde{W}$ can be expressed by

$$\begin{aligned}(w^1,w^2)_{W\hat{\otimes}\tilde{W}} &= (w^1_{11},w^2_{11})_{Y\hat{\otimes}\tilde{Y}} + \tilde{\rho}\cdot(w^1_{12},w^2_{12})_{Y\hat{\otimes}\tilde{Z}} \\ &\quad + \rho\cdot(w^1_{21},w^2_{21})_{Z\hat{\otimes}\tilde{Y}} + \rho\tilde{\rho}\cdot(w^1_{22},w^2_{22})_{Z\hat{\otimes}\tilde{Z}}\,,\end{aligned} \tag{2.7}$$

where $w^i = \begin{pmatrix} w^i_{11} & w^i_{12} \\ w^i_{21} & w^i_{22} \end{pmatrix}$, $i=1,2$; see [5].

Examples. In our illustrative examples we will make use of the usual Sobolew spaces

$W_2^k[a,b]$ and $W_2^{k,k}([a,b]\times[\tilde{a},\tilde{b}])$ endowed with the inner products

$$(x_1,x_2)_{W_2^k} = \sum_{i=0}^{k} \int_a^b D^i x_1(t)\cdot D^i x_2(t)\, dt \quad \text{and}$$

$$(y_1,y_2)_{W_2^{k,k}} = \sum_{i=0}^{k} \sum_{j=0}^{k} \int_a^b \int_{\tilde{a}}^{\tilde{b}} D^{i,j} y_1(t,\tilde{t})\cdot D^{i,j} y_2(t,\tilde{t})\, d\tilde{t}\, dt\,,$$

respectively. We point out, that the expected relations

$$W_2^{k,k}([a,b]\times[\tilde{a},\tilde{b}]) = W_2^k[a,b] \hat{\otimes} W_2^k[\tilde{a},\tilde{b}] \tag{2.8}$$

do really hold, see [11].

For $Z = \mathbb{R}^m$ and $\tilde{Z} = \mathbb{R}^{\tilde{m}}$ the tensor product $Z \hat{\otimes} \tilde{Z} = Z \otimes \tilde{Z}$ can be considered as the linear space of all $m\times\tilde{m}$-matrices, see [3].

The inner products $(z^1,z^2)_Z := \sum_{i=1}^{m} \omega_i z^1_i z^2_i$, $(\tilde{z}^1,\tilde{z}^2)_{\tilde{Z}} := \sum_{j=1}^{\tilde{m}} \tilde{\omega}_j \tilde{z}^1_j \tilde{z}^2_j$

lead to $(z^{o1},z^{o2})_{Z\otimes\tilde{Z}} = \sum_{i=1}^{m} \sum_{j=1}^{\tilde{m}} \omega_i\tilde{\omega}_j z^{o1}_{ij} z^{o2}_{ij}$.

Now, let us introduce the operators, which will play a role in our examples; and their tensor products.

For the second derivatives $D^2: W_2^2[a,b] \to W_2^0[a,b]: x \to D^2x$ and $\tilde{D}^2: W_2^2[\tilde{a},\tilde{b}] \to W_2^0[\tilde{a},\tilde{b}]: \tilde{x} \to D^2\tilde{x}$ the relation

$$D^{2,2} = D^2 \hat{\otimes} \tilde{D}^2 \tag{2.9}$$

holds, where $D^{2,2}: W_2^{2,2}([a,b]\times[\tilde{a},\tilde{b}]) \to W_2^{0,0}([a,b]\times[\tilde{a},\tilde{b}])$ is the mixed fourth partial derivative.

Further, for

$A : W_2^2[a,b] \to R^m: x \to (x(\tau_1),\ldots,x(\tau_m))^T$ and

$\tilde{A} : W_2^2[\tilde{a},\tilde{b}] \to R^{\tilde{m}}: \tilde{x} \to (\tilde{x}(\tilde{\tau}_1),\ldots,\tilde{x}(\tilde{\tau}_{\tilde{m}}))^T$ we get

$$A \hat{\otimes} \tilde{A}\, x^o = \begin{pmatrix} x^o(\tau_1,\tilde{\tau}_1) & \ldots & x^o(\tau_1,\tilde{\tau}_{\tilde{m}}) \\ \vdots & & \vdots \\ x^o(\tau_m,\tilde{\tau}_1) & \ldots & x^o(\tau_m,\tilde{\tau}_{\tilde{m}}) \end{pmatrix} \tag{2.10}$$

for $x^o \in W_2^{2,2}([a,b]\times[\tilde{a},\tilde{b}])$.

Finally,

$$D^2 \hat{\otimes} \tilde{A}\, x^o = (D^{2,0}x^o(.,\tilde{\tau}_1),\ldots,D^{2,0}x^o(.,\tilde{\tau}_{\tilde{m}}))^T. \tag{2.11}$$

Further explanations can be found in [3].

Cubic splines. In order to be complete we also give the definitions of the cubic spline spaces which figure in our examples.

The space of cubic splines with respect to the knot set $\{a=\tau_1<\ldots<\tau_m=b\}$ consists of all functions $s: [a,b] \to R$ with the properties

$s_i := s|_{[\tau_i,\tau_{i+1}]}$ is a cubic polynomial $(i=1,\ldots,m-1)$ and

$s \in C^2[a,b]$.

The space of natural cubic splines is defined by the additional conditions $D^2s(a) = D^2s(b) = 0$.

The space of bicubic splines with respect to the rectangular grid $\{a=\tau_1<\ldots<\tau_m\} \times \{\tilde{a}=\tilde{\tau}_1<\ldots<\tilde{\tau}_{\tilde{m}}=\tilde{b}\}$ consists of all functions $s: [a,b]\times[\tilde{a},\tilde{b}] \to R$ possessing the properties

$s|_{[\tau_i,\tau_{i+1}]\times[\tilde{\tau}_j,\tilde{\tau}_{j+1}]}$ is a bicubic polynomial $(i=1,\ldots,m-1;\ j=1,\ldots,\tilde{m}-1)$ and

$s \in C^{2,2}([a,b]\times[\tilde{a},\tilde{b}])$,

and it turns out to be the tensor product of the univariate cubic splines with respect to the individual knot sets of the grid.

Natural bicubic splines fulfil the additional conditions (see [1], [3])

$D^{2,0}s(a,\tilde{\tau}_j) = D^{2,0}s(b,\tilde{\tau}_j) = 0 \quad (j=1,\ldots,\tilde{m})$,

$D^{0,2}s(\tau_i,\tilde{a}) = D^{0,2}s(\tau_i,\tilde{b}) = 0 \quad (i=1,\ldots,m)$ and

$D^{2,2}s(a,\tilde{a}) = D^{2,2}s(b,\tilde{a}) = D^{2,2}s(a,\tilde{b}) = D^{2,2}s(b,\tilde{b}) = 0.$

3. Extremal Properties of Interpolating Tensor Product Splines

We start with a concise description of the variatonal approach to abstract interpolating splines in Hilbert spaces. Therefore we make use of the following assumptions:

$$X,\ Y,\ Z \quad \text{Hilbert spaces of real functions,}$$
$$T: X \to Y, \quad A: X \to Z \quad \text{continuous surjective linear operators,} \tag{3.1}$$

$$\ker T \ + \ \ker A \quad \text{closed in } X, \tag{3.2}$$

$$\ker T \cap \ker A \ = \ \{0\}. \tag{3.3}$$

The corresponding abstract spline space $Sp(T,A)$ in X is defined by

$$Sp(T,A) := \{x \in X: \ (Tx,Th)_Y = 0 \ \text{ for all } \ h \in \ker A\}. \tag{3.4}$$

The interpolating spline $s = s(T,A,z)$ to $z \in Z$ with respect to A and T is obtained as the solution of the extremal problem

$$\text{minimize } \|Tx\|_Y^2 \qquad \text{over} \quad x \in A^{-1}z := \{x \in X: \ Ax = z\}, \tag{3.5}$$

which is uniquely solvable. Here the assumption (3.2) is necessary for the existence, while the assumption (3.3) assures the uniqueness of the interpolating spline, see [2, Theorem 4.1].
With regard to our considerations of the tensor product scheme we introduce a slightly modified, but equivalent formulation of problem (3.5). To this end let U be an additional Hilbert space of real functions and $B: X \to U$ a continuous linear operator possessing the properties

$$\ker A \subset \ker B \quad \text{and} \quad \ker T \cap \ker B = \{0\}. \tag{3.6}$$

The problem (3.5) is equivalent to the problem

$$\text{minimize } \|Sx\|_W^2 \qquad \text{over} \quad x \in A^{-1}z, \tag{3.7}$$

where $W := Y \times U$ (see (2.5)) and the operator $S: X \to W$ is defined by $Sx := (Tx, Bx)$. Under the formulated assumptions the operator S has the properties

$$\ker S = \{0\} \quad \text{and} \quad \text{im } S \quad \text{closed in } W. \tag{3.8}$$

Therefore the unique solution $s = s(T,A,z) = s(S,A,z)$ of the equivalent problems (3.5) and (3.7) is completly characterized by the conditions

$$As = z \quad \text{and} \quad (Ss,Sh)_W = 0 \quad \text{for all} \quad h \in \ker A, \tag{3.9}$$

which means, that $s \in Sp(T,A) = Sp(S,A)$, see [2, Theorem 4.1].

In most applications the (finite-dimensional) spline space Sp(S,A) is known in advance, and the extremal problem (3.7) is handled in a constructive way considering the interpolation problem (3.9), which leads to a linear interpolation operator

$$I_{Sp(S,A)}: Z \to Sp(S,A): z \to s(S,A,z). \tag{3.10}$$

In this context it is referred to problem (3.5) as the extremal property of interpolating splines.

It is usual to construct bivariate interpolating splines by the formation of the tensor product of two interpolation operators as defined in (3.10) (e.g. [1], [3], [16]). Here, we are interested in the extremal properties of such tensor product splines.
Let a second set of Hilbert spaces $\tilde{X}$, $\tilde{Y}$, $\tilde{Z}$, $\tilde{U}$ and operators $\tilde{A}$, $\tilde{T}$, $\tilde{B}$ be given, such that the assumptions (3.1) - (3.3) and (3.6) hold. The space $\tilde{W}$ and the operator $\tilde{S}$ are defined in an analogous manner as above. A first attempt consists in the consideration of the following problem:

$$\text{minimize } \|T\hat{\otimes}\tilde{T}\, x^0\|^2_{Y\hat{\otimes}\tilde{Y}} \quad \text{over } x^0 \in (A\hat{\otimes}\tilde{A})^{-1}\, z^0. \tag{3.11}$$

In general this problem has more than one solution, because $\ker T\hat{\otimes}\tilde{T} \cap \ker A\hat{\otimes}\tilde{A} \neq \{0\}$. Therefore we deal with the problem

$$\text{minimize } \|S\hat{\otimes}\tilde{S}\, x^0\|^2_{W\hat{\otimes}\tilde{W}} \quad \text{over } x^0 \in (A\hat{\otimes}\tilde{A})^{-1}\, z^0, \tag{3.12}$$

which is equivalent to

$$\text{minimize } \|T\hat{\otimes}\tilde{T}\, x^0\|^2_{Y\hat{\otimes}\tilde{Y}} + \tilde{\rho}\|T\hat{\otimes}\tilde{B}\, x^0\|^2_{Y\hat{\otimes}\tilde{U}} + \rho\|B\hat{\otimes}\tilde{T}\, x^0\|^2_{U\hat{\otimes}\tilde{Y}}$$
$$\text{over } x^0 \in (A\hat{\otimes}\tilde{A})^{-1}\, z^0.$$

Owing to $\ker S\hat{\otimes}\tilde{S} = \{0\}$ and to the closedness of $\operatorname{im} S\hat{\otimes}\tilde{S}$ the problem (3.12) is uniquely solvable and the solution $s^0 = s^0(S\hat{\otimes}\tilde{S}, A\hat{\otimes}\tilde{A}, z^0)$ is characterized by

$$A\hat{\otimes}\tilde{A}\, s^0 = z^0 \quad \text{and} \quad s^0 \in Sp(S\hat{\otimes}\tilde{S}, A\hat{\otimes}\tilde{A}), \tag{3.13}$$

which leads to the interpolation operator

$$I_{Sp(S\hat{\otimes}\tilde{S}, A\hat{\otimes}\tilde{A})}: Z\hat{\otimes}\tilde{Z} \to Sp(S\hat{\otimes}\tilde{S}, A\hat{\otimes}\tilde{A}), \tag{3.14}$$

compare (3.9), (3.10).
We get the following result.

<u>Theorem 1:</u> $I_{Sp(S\hat{\otimes}\tilde{S}, A\hat{\otimes}\tilde{A})} = I_{Sp(S,A)} \hat{\otimes} I_{Sp(\tilde{S},\tilde{A})}$.

Proof: We only sketch the main ideas of the proof. Fore more detailed considerations using adjoint operators we refer to [5].

At first we show that

$$s^{o}(S\hat{\otimes}\tilde{S}, A\hat{\otimes}\tilde{A}, z\otimes\tilde{z}) = s(S,A,z) \otimes \tilde{s}(\tilde{S},\tilde{A},\tilde{z}) \quad \text{for } z \in Z,\ \tilde{z} \in \tilde{Z}.$$

Indeed, with $s = s(S,A,z)$, $\tilde{s} = \tilde{s}(\tilde{S},\tilde{A},\tilde{z})$ we get

$$A \hat{\otimes} \tilde{A}\ (s\otimes\tilde{s}) = As \otimes \tilde{A}\tilde{s} = z \otimes \tilde{z} \quad \text{and}$$

$$(S\hat{\otimes}\tilde{S}\ (s\otimes\tilde{s}), S\hat{\otimes}\tilde{S}\ (h\otimes\tilde{h}))_{W\hat{\otimes}\tilde{W}} = (Ss,Sh)_W \cdot (\tilde{S}\tilde{s},\tilde{S}\tilde{h})_{\tilde{W}} = 0$$

for $h \in \ker A$ or $\tilde{h} \in \ker \tilde{A}$, which induces

$$(S\hat{\otimes}\tilde{S}\ (s\otimes\tilde{s}), S\hat{\otimes}\tilde{S}\ h^{o})_{W\hat{\otimes}\tilde{W}} = 0 \quad \text{for all } h^{o} \in \ker A\hat{\otimes}\tilde{A}.$$

Therefore, $s^{o} = s \otimes \tilde{s}$ fulfils the conditions (3.13).
Now, the result is obtained by linear and continuous extension.

This result shows, that the constructive tensor product scheme of the spline interpolation operators $I_{Sp(S,A)}$ and $I_{Sp(\tilde{S},\tilde{A})}$ is uniquely characterized by the extremal property formulated in problem (3.12). We point out, that the objective functional in problem (3.12) can be modified by changing the operators B, $\tilde{B}$, the corresponding image spaces U, $\tilde{U}$ and the inner products in U, $\tilde{U}$ with respect to the properties (3.6) without affecting the spline space and the interpolating tensor product spline.

Example 3.1: Natural bicubic interpolating spline [1], [6], [8], [16].
We determine $X = W_2^2[a,b]$, $Y = W_2^0[a,b]$, $Z = R^m$,

$$T = D^2, \quad A: X \to Z: x \to (x(\tau_1), \dots, x(\tau_m))^T,$$

$$U = Z, \quad (u,v)_U = \sum_{i=1}^{m} \omega_i u_i v_i \quad (\omega_i > 0), \quad B = A.$$

It is well known, that in this case the solution $s = s(T,A,z)$ of the resulting problem

$$\text{minimize} \int_a^b [D^2 x(t)]^2 dt \quad \text{over} \quad x \in \{x \in W_2^2[a,b]: \; x(\tau_i) = z_i \ (i=1,\dots,m)\} \tag{3.15}$$

is the natural cubic interpolating spline. Let $\tilde{X}$, ..., $\tilde{U}$ be defined analogously. The corresponding natural bicubic interpolating spline turns out to be the unique solution of the extremal problem

$$\begin{aligned} \text{minimize } J(x^{o}) := & \int_a^b \int_{\tilde{a}}^{\tilde{b}} [D^{2,2} x^{o}(t,\tilde{t})]^2 d\tilde{t}dt + \\ & \rho \cdot \sum_{i=1}^{m} \omega_i \cdot \int_{\tilde{a}}^{\tilde{b}} [D^{0,2} x^{o}(\tau_i, t)]^2 d\tilde{t} + \\ & \tilde{\rho} \cdot \sum_{j=1}^{\tilde{m}} \tilde{\omega}_j \cdot \int_a^b [D^{2,0} x^{o}(t,\tilde{\tau}_j)]^2 dt \end{aligned} \tag{3.16}$$

$$\text{over} \quad x^{o} \in \{x^{o} \in W_2^{2,2}([a,b]\times[\tilde{a},\tilde{b}]): x^{o}(\tau_i,\tilde{\tau}_j) = z^{o}_{ij}\}.$$

Of course, the functional J can be modified in many ways; e.g., by choosing $U = \mathbb{R}^2$, $B: X \to U: x \to (x(\tau_1), x(\tau_m))^T$ and $(u,v)_U = \omega_1 u_1 v_1 + \omega_m u_2 v_2$ ($\tilde{U}$, $\tilde{B}$ analogously).

In this case the objective functional

$$J(x^0) = \int_a^b \int_{\tilde{a}}^{\tilde{b}} [D^{2,2}x^0(t,\tilde{t})]^2 d\tilde{t}dt + \rho\omega_1 \cdot \int_{\tilde{a}}^{\tilde{b}} [D^{0,2}x^0(\tau_1,\tilde{t})]^2 d\tilde{t}$$
$$+ \rho\omega_m \cdot \int_{\tilde{a}}^{\tilde{b}} [D^{0,2}x^0(\tau_m,\tilde{t})]^2 d\tilde{t} + \tilde{\rho}\tilde{\omega}_1 \cdot \int_a^b [D^{2,0}x^0(t,\tilde{\tau}_1)]^2 dt$$
$$+ \tilde{\rho}\tilde{\omega}_{\tilde{m}} \cdot \int_a^b [D^{2,0}x^0(t,\tilde{\tau}_{\tilde{m}})]^2 dt$$

yields for problem (3.16) resulting in the same solution.

Example 3.2: Complete bicubic interpolating spline [1], [2], [7], [16].

X, Y, T, U, B are defined as before, but

$Z = \mathbb{R}^{m+2}$: $z = (z_0, \ldots, z_{m+1})^T$, $A: X \to Z: x \to (Dx(\tau_1), x(\tau_1), \ldots, Dx(\tau_m))^T$.

The solution of the resulting problem

$$\text{minimize} \int_a^b [D^2x(t)]^2 dt \quad \text{over} \quad x \in \{x \in W_2^2[a,b]: x(\tau_i)=z_i \ (i=1,\ldots,m),\ Dx(\tau_1)=z_0,\ Dx(\tau_m)=z_{m+1}\}, \qquad (3.17)$$

is a cubic spline called complete interpolating spline.

Let $\tilde{X}, \ldots, \tilde{U}$ be defined analogously. The corresponding complete bicubic interpolating spline appears as the unique solution of

$$\begin{aligned}
&\text{minimize } J(x^0)\\
&\text{over } x^0 \in \{x^0 \in W_2^{2,2}([a,b]\times[\tilde{a},\tilde{b}]): x^0(\tau_i,\tilde{\tau}_j)=z^0_{ij},\\
&\quad D^{0,1}x^0(\tau_i,\tilde{\tau}_1)=z^0_{i0};\ D^{0,1}x^0(\tau_i,\tilde{\tau}_{\tilde{m}})=z^0_{i,\tilde{m}+1} \ (i=1,\ldots,m),\\
&\quad D^{1,0}x^0(\tau_1,\tilde{\tau}_j)=z^0_{0j},\ D^{1,0}x^0(\tau_m,\tilde{\tau}_j)=z^0_{m+1,j} \ (j=1,\ldots,\tilde{m}), \qquad (3.18)\\
&\quad D^{1,1}x^0(\tau_1,\tilde{\tau}_1)=z^0_{00},\ D^{1,1}x^0(\tau_m,\tilde{\tau}_m)=z^0_{m+1,\tilde{m}+1},\\
&\quad D^{1,1}x^0(\tau_m,\tilde{\tau}_1)=z^0_{m+1,0},\ D^{1,1}x^0(\tau_1,\tilde{\tau}_{\tilde{m}})=z^0_{0,\tilde{m}+1}\}.
\end{aligned}$$

Here the introduction of the additional operators B, $\tilde{B}$ results in a simpler formulation of the objective functional than the choice $B = A$, $\tilde{B} = \tilde{A}$ would give, compare example 4.2.

Of course, using the abstract approach sketched in this paper, it is possible to form the tensor product scheme of arbitrary combinations of all special types of interpolating splines considered in [2].

For example, in [1], [6], [7], [8] it is dealt with tensor products of polynomial splines of odd degree, while in [1], [17] also L-splines are considered. All these investigations turn out to be particular cases of the abstract tensor product scheme considered above.

4. Smoothing Tensor Product Splines

In many applications the data $z \in Z$ are measurements affected by errors, and therefore the strong interpolation condition $Ax = z$ is not adequate. In these cases it is usual to combine the extremal (smoothing) properties of splines with the idea of fitting in the sense of least squares. On the basis of the assumptions (3.1)-(3.3), the smoothing spline $\sigma_\rho = \sigma_\rho(T,A,z)$ to $z \in Z$ with respect to T, A, ρ is defined as the solution of the problem

$$\text{minimize } \|Tx\|_Y^2 + \rho\|Ax - z\|_Z^2 \quad \text{over } x \in X, \tag{4.1}$$

which is uniquely solvable.

Here, $\rho>0$ is a given smoothing parameter balancing goodness of fit against smoothness. For a nice discussion of the influence of the parameter ρ on the smoothing spline we refer to [12], [13].

The problem (4.1) is equivalent to

$$\text{minimize } \|Sx - w\|_W^2 \quad \text{over } x \in X, \tag{4.2}$$

where $W := Y \times Z$ (see (2.5)), $w := (0,z) \in W$ and the operator $S\colon X \to W$ is defined by $Sx := (Tx,Ax)$, compare the formulation of problem (3.7). We use this equivalence to introduce the shorter notations $\sigma = \sigma(S,w) := \sigma_\rho(T,A,z)$. Analogous arguments as in the considerations of problem (3.7) assure the existence and uniqueness of $\sigma = \sigma(S,w)$, which is characterized by the conditions

$$\sigma \in X \quad \text{and} \quad (S\sigma - w, Sx)_W = 0 \quad \text{for all } x \in X. \tag{4.3}$$

It is also easy to realize that σ is a spline, which means that $\sigma \in Sp(T,A) = Sp(S,A)$, see [2, Theorem 4.2].

Since every $x \in X$ can be written in the form

$$x = s(S,A,Ax) + (\, x - s(S,A,Ax)\,),$$

the condition (4.3) is equivalent to

$$(S\sigma-w,Sx)_W = 0 \quad \text{for all } x \in Sp(S,A) \text{ and for all } x \in \ker A,$$

which can be reduced to

$$\sigma \in Sp(S,A) \quad \text{and} \quad (S\sigma-w,Sx)_W=0 \quad \text{for all } x \in Sp(S,A). \tag{4.4}$$

In most applications the spline space Sp(S,A) is finite-dimensional and known in advance. Given a basis of Sp(S,A), it is possible to compute the smoothing spline σ by solving the "normal equations" (4.4), see, e.g., [9]. One is led to a linear smoothing operator

$$G_{Sp(S,A)}: Z \to Sp(S,A): z \to \sigma(S,(0,z)). \qquad (4.5)$$

Much effort has been done to find appropriate objective functionals for bivariate smoothing splines in order to reach the main advantage of a separating solution. As in the case of interpolating tensor product splines this property leads to highly efficient numerical algorithms for computing bivariate smoothing splines. In the literature there can be found some isolated results for particular cases, mainly for polynomial splines, see [4], [6], [7], [16], [17]. Here we describe the abstract approach to smoothing tensor product splines, which includes all the investigations mentioned above as special cases.

Let a second set of Hilbert spaces $\tilde{X}$, $\tilde{Y}$, $\tilde{Z}$ and operators $\tilde{A}$, $\tilde{T}$ be given, such that the assumptions (3.1) -(3.3) are fulfilled, and define the space $\tilde{W}$ and the operator $\tilde{S}$ analogously as W, S.
We deal with the problem

$$\text{minimize } \|S\hat{\otimes}\tilde{S}\, x^0 - w^0\|^2_{W\hat{\otimes}\tilde{W}} \quad \text{over } x^0 \in X \hat{\otimes} \tilde{X}, \qquad (4.6)$$

where $w^0 = \begin{pmatrix} 0 & 0 \\ 0 & z^0 \end{pmatrix} \in W \hat{\otimes} \tilde{W}$ is given.
The problem (4.6) also reads as follows:

$$\begin{aligned} \text{minimize } & \|T\hat{\otimes}\tilde{T}\, x^0\|^2_{Y\hat{\otimes}\tilde{Y}} + \tilde{\rho}\|T\hat{\otimes}\tilde{A}\, x^0\|^2_{Y\hat{\otimes}\tilde{Z}} + \rho\|A\hat{\otimes}\tilde{T}\, x^0\|^2_{Z\hat{\otimes}\tilde{Y}} \\ & + \rho\tilde{\rho}\|A\hat{\otimes}\tilde{A}\, x^0 - z^0\|^2_{Z\hat{\otimes}\tilde{Z}} \quad \text{over } x^0 \in X \hat{\otimes} \tilde{X}. \end{aligned} \qquad (4.7)$$

Using the same arguments as in 3., it can be shown that the problem (4.6) is uniquely solvable and that its solution $\sigma^0 = \sigma^0(S\hat{\otimes}\tilde{S}, w^0)$ is characterized by $\sigma^0 \in X \hat{\otimes} \tilde{X}$ and

$$(S\hat{\otimes}\tilde{S}\, \sigma^0 - w^0, S\hat{\otimes}\tilde{S}\, x^0)_{W\hat{\otimes}\tilde{W}} = 0 \quad \text{for all } x^0 \in X \hat{\otimes} \tilde{X}. \qquad (4.8)$$

Therefore we are led to a linear smoothing operator,

$$G_{Sp(S\hat{\otimes}\tilde{S}, A\hat{\otimes}\tilde{A})}: Z\hat{\otimes}\tilde{Z} \to X\hat{\otimes}\tilde{X}: z^0 \to \sigma^0(S\hat{\otimes}\tilde{S}, w^0), \qquad (4.9)$$

and get the following result:

<u>Theorem 2:</u> $\quad G_{Sp(S\hat{\otimes}\tilde{S}, A\hat{\otimes}\tilde{A})} = G_{Sp(S,A)} \hat{\otimes} G_{Sp(\tilde{S},\tilde{A})}$.

Proof: We give a concise sketch of the proof, which follows the same lines as the proof of Theorem 1. For more detailed considerations we refer to [5] again. At first , we show that
$\sigma^0(S\hat{\otimes}\tilde{S}, w\otimes\tilde{w}) = \sigma(S,w) \otimes \tilde{\sigma}(\tilde{S},\tilde{w})$ for $w=(0,z) \in W$, $\tilde{w}=(0,\tilde{z}) \in \tilde{W}$:
With $\sigma=\sigma(S,w)$, $\tilde{\sigma}=\tilde{\sigma}(\tilde{S},\tilde{w})$ we get

$$(S\hat{\otimes}\tilde{S}\, (\sigma\otimes\tilde{\sigma}), S\hat{\otimes}\tilde{S}\, (x\otimes\tilde{x}))_{W\hat{\otimes}\tilde{W}} = (S\sigma, Sx)_W \cdot (\tilde{S}\tilde{\sigma}, \tilde{S}\tilde{x})_{\tilde{W}}$$

$= (w,Sx)_W \cdot (\tilde{w},\tilde{S}\tilde{x})_{\tilde{W}} = (w\otimes\tilde{w}, S\hat{\otimes}\tilde{S}\ (x\otimes\tilde{x}))_{W\hat{\otimes}\tilde{W}}$ for all $x \in X$, $\tilde{x} \in \tilde{X}$.

Then, $(S\hat{\otimes}\tilde{S}\ (\sigma\otimes\tilde{\sigma}), S\hat{\otimes}\tilde{S}\ x^o)_{W\hat{\otimes}\tilde{W}} = (w\otimes\tilde{w}, S\hat{\otimes}\tilde{S}x^o)_{W\hat{\otimes}\tilde{W}}$ for all $x^o \in X\hat{\otimes}\tilde{X}$

follows by linear and continuous extension, and with respect to (4.8) the assertion is proved.

Now the result of Theorem 2 is obtained by linear and continuous extension.

That means, that the solution operator of problem (4.6) is obtained as the constructive tensor product of the two smoothing operators $G_{Sp(S,A)}$ and $G_{Sp(\tilde{S},\tilde{A})}$. In practice, this result is expressed in the fact, that the computation separates into the solution of a number of problems of low dimension. Therefore, the main advantage of interpolating tensor product splines is reproduced for smoothing splines. For computational details we refer to [3], [5] and to the investigations of special cases in [6], [7] and [16].

Example 4.1: Natural bicubic smoothing spline [6], [15], [16].

On the basis of the assumptions of example 3.1 the natural cubic smoothing spline is the solution of the following problem:

$$\text{minimize} \int_a^b [D^2x(t)]^2 dt + \rho \cdot \sum_{i=1}^{m} \omega_i [x(\tau_i) - z_i]^2 \quad \text{over } x \in W_2^2[a,b]. \tag{4.10}$$

The corresponding natural bicubic smoothing spline solves the problem:

$$\text{minimize} \quad J(x^o) + \rho\tilde{\rho} \sum_{i=1}^{m} \sum_{j=1}^{m} [x^o(\tau_i,\tilde{\tau}_j) - z_{ij}^o]^2 \cdot \omega_i \tilde{\omega}_j \qquad \text{over} \quad x^o \in W_2^{2,2}([a,b]\times[\tilde{a},\tilde{b}]), \tag{4.11}$$

compare (3.16).

Example 4.2: Complete bicubic smoothing spline [7].

In this case, see example 3.2, the complete cubic smoothing spline is obtained as the solution of the problem

$$\text{minimize} \int_a^b [D^2x(t)]^2 dt + \rho\{\omega_0[Dx(\tau_1)-z_0]^2 + \sum_{i=1}^{m} \omega_i [x(\tau_i)-z_i]^2 + \omega_{m+1}[Dx(\tau_m)-z_{m+1}]^2\} \qquad \text{over} \quad x \in W_2^2[a,b]. \tag{4.12}$$

The solution of this problem is an arbitrary cubic spline, while in the first example only natural cubic splines occur. The difference consists in the fact, that in the second example also data for the derivatives at the boundaries enter into the fit.

The corresponding tensor product scheme, the complete bicubic smoothing spline, appears as the unique solution of the problem

$$\text{minimize } J^0(x^0) \quad \text{over } x^0 \in W_2^{2,2}([a,b]\times[\tilde a,\tilde b]), \tag{4.13}$$

where the objective functional J^0 is given by

$$\begin{aligned} J^0(x^0) = {} & \int_a^b \int_{\tilde a}^{\tilde b} [D^{2,2}x^0(t,\tilde t)]^2 d\tilde t\, dt \\ & + \rho\{\omega_0 \cdot \int_{\tilde a}^{\tilde b}[D^{1,2}x^0(\tau_1,\tilde t)]^2 d\tilde t + \sum_{i=1}^{m} \omega_i \cdot \int_{\tilde a}^{\tilde b}[D^{0,2}x^0(\tau_i,\tilde t)]^2 d\tilde t \\ & \quad + \omega_{m+1} \cdot \int_{\tilde a}^{\tilde b}[D^{1,2}x^0(\tau_m,\tilde t)]^2 d\tilde t\} \\ & + \tilde\rho\{\tilde\omega_0 \cdot \int_a^b [D^{2,1}x^0(t,\tau_1)]^2 dt + \sum_{j=1}^{\tilde m} \tilde\omega_j \cdot \int_a^b [D^{2,0}x^0(t,\tilde\tau_j)]^2 dt \\ & \quad + \tilde\omega_{\tilde m+1} \cdot \int_a^b [D^{2,1}x^0(t,\tilde\tau_{\tilde m})]^2 dt\} \\ & + \rho\tilde\rho\{\omega_0\tilde\omega_0[D^{1,1}x^0(\tau_1,\tilde\tau_1)-z^0_{00}]^2 + \omega_{m+1}\tilde\omega_{\tilde m+1}[D^{1,1}x^0(\tau_m,\tilde\tau_{\tilde m})-z^0_{m+1,\tilde m+1}]^2 \\ & \quad + \omega_{m+1}\tilde\omega_0[D^{1,1}x^0(\tau_m,\tilde\tau_1)-z^0_{m+1,0}]^2 + \omega_0\tilde\omega_{\tilde m+1}[D^{1,1}x^0(\tau_1,\tilde\tau_{\tilde m})-z^0_{0,\tilde m+1}]^2 \\ & \quad + \sum_{i=1}^{m}(\omega_i\tilde\omega_0[D^{0,1}x^0(\tau_i,\tilde\tau_1)-z^0_{i0}]^2 + \omega_i\tilde\omega_{\tilde m+1}[D^{0,1}x^0(\tau_i,\tilde\tau_{\tilde m})-z^0_{i,\tilde m+1}]^2) \\ & \quad + \sum_{j=1}^{\tilde m}(\omega_0\tilde\omega_j[D^{1,0}x^0(\tau_1,\tilde\tau_j)-z^0_{0j}]^2 + \omega_{m+1}\tilde\omega_j[D^{1,0}x^0(\tau_m,\tilde\tau_j)-z^0_{m+1,j}]^2) \\ & \quad + \sum_{i=1}^{m}\sum_{j=1}^{\tilde m} \omega_i\tilde\omega_j[x^0(\tau_i,\tilde\tau_j)-z^0_{ij}]^2\}. \end{aligned}$$

We want to point out, that the proofs for the particular cases in [6], [7] and [16] are given by intricate calculations using the special definitions of the objective functionals. In this paper the results are derived as direct consequences from the abstract approach and from the well-known results for the related univariate splines. Of course, it is possible to form the tensor product scheme of arbitrary combinations of all types of smoothing splines considered in [2].

Acknowledgement: The third named author is indebted to N.N. Pavlov and V.V. Vershinin for helpful discussions during his stay in Novosibirsk in October 1988.

References

[1] Ahlberg,J.H., E.N.Nilson, J.L.Walsh: The theory of splines and their applications. Academic Press, New York-London 1967.

[2] Böhmer,K.: Spline-Funktionen. B.G.Teubner,Stuttgart 1974.

[3] de Boor,C.: A practical guide to splines. Springer-Verlag, New York-Heidelberg-Berlin 1978.

[4] Dierckx,P.: A fast algorithm for smoothing data on a rectangular grid while using spline functions. SIAM J. Numer.Anal. 19(1982),1286-1304.

[5] Ewald,S.: Übertragung von Approximationsverfahren auf Tensorprodukträume. Diplomarbeit, TU Dresden 1988.

[6] Hu,C.L., L.L.Schumaker: Bivariate natural spline smoothing. In: Delay equations, approximation and applications (Mannheim 1984), ISNM 74. Birkhäuser-Verlag, Basel 1985, 165-179.

[7] Hu,C.L., L.L.Schumaker: Complete spline smoothing. Numer.Math. 49(1986),1-10.

[8] Imamov,A.: About some properties of multivariate splines (in Russian). Vycislitelnye sistemy (Novosibirsk) 65(1975),68-73.

[9] Lyche,T., L.L.Schumaker: Computation of smoothing and interpolating natural splines via local bases. SIAM J. Numer.Anal. 10(1973),1027-1038.

[10] Murray,F.H., J. von Neumann: On rings of operators. Ann. of Math. 37(1936),116-229.
= Collected works of J. von Neumann, III. Pergamon Press, 1961, 6-119.

[11] Reed,M., B.Simon: Methods of modern mathematical physics I. Academic Press, New York-London 1972.

[12] Reinsch,C.H.: Smoothing by spline functions. Numer.Math. 10 (1967),177-183.

[13] Reinsch,C.H.: Smoothing by spline functions II. Numer.Math. 16 (1971),451-454.

[14] Sard,A.: Linear approximation. Mathematical Surveys 9, AMS, Providence 1963.

[15] Savjalov,J.S., B.I.Kvasov, V.L.Miroshnichenko: Methods of spline functions (in Russian). Nauka, Moscow 1980.

[16] Savjalov,J.S., A.Imamov: Algorithm with separation for the solution of the problem of smoothing by multivariate spline functions (in Russian). Čislennye metody mechaniki složnoi sredy, Novosibirsk 1976, 52-61.

[17] Vershinin,V.V., J.S.Savjalov, N.N.Pavlov: Extremal properties of splines and the problem of smoothing (in Russian). Nauka, Novosibirsk 1988.

SPLINES OF FINITE ELEMENT TYPE FOR ENCLOSING METHODS OF SECOND ORDER DIFFERENTIAL EQUATIONS

Andreas Felgenhauer [1]

0. Introduction

For the model problem

$$\begin{aligned} -\Delta U &= g(x, u) \quad & x \in \Omega \subset R^N \\ u &= 0 & x \in \partial\Omega \end{aligned} \tag{1}$$

enclosing methods have been developed in a series of publications (e. g. [5] , [1] , [4]). In the one-dimensional case N = 1 efficient numerical algorithms for piecewise polynomial sub- and supersolutions were constructed. New principal difficulties arise if the number of dimensions is greater than one. They are based on the fact that for $N > 1$ exact solution of (1) with arbitrary polynomial right-hand side is impossible even if Ω is a finite element.

In the first part of this paper, we give some arguments how to overcome this barrier. In the second we propose concrete algorithms based on Bogner-Fox-Schmit elements to construct enclosures for (1) in the case that Ω can be triangulated by rectangles (N = 2).

1. Monotone enclosures

We treat (1) as an operator equation

$$Au = Lu - Gu = f, \quad u \in V \tag{2}$$

in the dual Banach space V_0^* where $V_0 = H_0^1(\Omega) \subset V \subset H^1(\Omega)$ and $f \in L^\infty(\Omega)$. L is supposed to be a linear operator and G may be nonlinear.

In $V_0 \subset V \subset L^2(\Omega)$ we consider the partial order relation $u \leq v$ defined by "pointwise almost everywhere". Then, the image space V_0^* is partial ordered by the dual of the positive cone in V_0. This definition is consistent with the imbedding

$$V \subset L^2(\Omega) \subset V_0^* \tag{3}$$

because $(L^2(\Omega), \leq)$ is a Banach lattice.

<u>Theorem 1:</u> Let L and G be differential operators on Ω of at most second order and A be strict monotone, i. e.

$$\langle Au - Av, u-v \rangle > 0 \tag{4}$$

for all $u \in V$, $u-v \in V_0$ $u \neq v$. Then A is inverse-monotone in the following sense.

---

1) TU "Otto von Guericke", Sektion Mathematik
DDR-3032 Magdeburg, Boleslaw-Bierut-Platz 5

If

$$Au \leq Av \quad \text{in} \quad V_o^* \tag{5}$$

$$u \leq v \quad \text{on} \quad \partial\Omega \tag{6}$$

then

$$u \leq v \quad \text{in} \quad V \;\blacksquare \tag{7}$$

The proof immediately follows from Glashoff, Werner [4], the required "L-property" is shown in Ladyshenskaya Ural'ceva [6] ∎

We aim at enclosing solutions of (2) based on Theorem 1 by constructing sub- and supersolutions $\underline{u}_h, \overline{u}_h \in V_h \subset V$

$$A\underline{u}_h = \underline{f}_h \leq f \leq \overline{f}_h = A\overline{u}_h \tag{8}$$

where V_h is a finite dimensional space of piecewise polynomial trial functions.

In the following we suppose, $G = 0$, $A = L$. The case $G \neq 0$ may be considered iterative.

Based on (8) we have to choose a suitable set $F_h \subset V_o^*$ and to construct an enclosure (11) for the right-hand side f. In a second step we have to solve

$$Lu_h = f_h \qquad f_h \in F_h \subset V_o^* \tag{9}$$

exactly. ($f_h = \underline{f}_h$ or $f_h = \overline{f}_h$). In the one dimensional case and for a special operator L, F_h may be chosen piecewise polynomial. But in the multidimensional case the equation (9) for arbitrary piecewise polynomial f_h, cannot be solved exactly even in the most simple case of the Laplacian $L = -\Delta$.

Therefore, we reverse the procedure and set

$$F_h = LV_h \; . \tag{10}$$

Now the difficulties consist in construction of a suitable enclosure

$$\underline{f}_h \leq f \leq \overline{f}_h , \qquad \underline{f}_h , \quad \overline{f}_h \in F_h \; . \tag{11}$$

For that reason we introduce a further finite dimensional space $\tilde{F}_h \subset F_h$, for instance the space of piecewise constants.

Instead of (11) we enclose

$$\underset{\sim}{f}_h \leq f \leq \tilde{f}_h , \qquad \underset{\sim}{f}_h , \quad \tilde{f}_h \in \tilde{F}_h \tag{12}$$

and in a second step we solve

$$\underline{f}_h \leq \underset{\sim}{f}_h , \quad \overline{f}_h \leq \underset{\sim}{\tilde{f}}_h , \quad \overline{f}_h \in F_h . \tag{13}$$

2. Regularity of trial functions

We discuss the inequality (11) in the sense of V_0^* under the assumption that V_h in (10) is a space of finite element functions.

Therefore, we suppose a triangulation $\mathcal{T}_h$ of the domain

$$\Omega = \bigcup_{K \in \mathcal{T}_h} K. \tag{14}$$

(We use the notation of Ciarlet [3]). Let L_K, f_K, u_K ... be the restriction of L, f_h, u_h ... to K, respectively, and

$$\mathcal{T}_h' = \{K' \ / \text{There exist } K_1 \neq K_2 \in \mathcal{T}_h \text{ with } K' = K_1 \cap K_2 \neq \emptyset\}$$

the net of the triangulation. For $u_h \in V_h$ we suppose at least

$$u_h \in C\ (\Omega),\ u_K \in C^1\ (K) \cap W^{2,\infty}\ (K).$$

Thus, an inequality

$$f_h = Lu_h \geq f$$

is equivalent to

$$f_K = L_K\, u_h \geq f, \quad \text{a. e.} \quad \text{in } K \text{ for all } K \in \mathcal{T}_h , \tag{15}$$

and

$$l_{K'}\, u_h \geq 0 \quad , \quad \text{a. e. in } K' \text{ for all } K' \in \mathcal{T}_h'. \tag{16}$$

Here $l_{K'}$ is a trace operator which arises from integration by parts of L_K on all (N-1)-dimensional components of $\mathcal{T}_h'$. For the special operator $Lu = -\Delta u + c\ (\ .\ )\ u$ we have

$$l_{K'}\, u_h = \frac{\partial u_h}{\partial n_1}\Big|_{K_1} + \frac{\partial u_h}{\partial n_2}\Big|_{K_2} \tag{17}$$

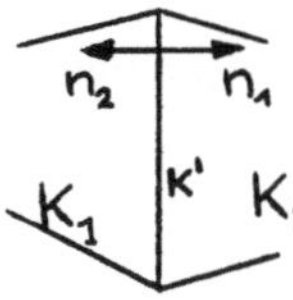

i. e. super-(sub-) solutions have to be concave (convex) in normal direction on K'.
Thus, for good approximation, a regularity condition

$$l_{K'}\, u_h = 0 \qquad \text{a. e. in } K' \text{ for all } K' \in \mathcal{T}_h', u_h \in V_h \tag{reg 1}$$

turns out to be efficient. In case (17), (reg 1) means the continuity of normal derivatives.
A sufficient condition for (reg 1) is

$$u_h \in C^1\ (\overline{\Omega}) \qquad \text{for all } u_h \in V_h , \tag{reg 2}$$

i. e. the f. e. space is of class $\mathcal{C}^1$.

3. An example for semiregular trial function (N = 2)

We restrict our consideration to the case of $L = -\Delta$.
Then (17) is fulfilled. The continuity of $\frac{\partial u_h}{\partial n}$ on K' may be guaranteed by the continuity of degrees of freedom of u_h on a set which is unisolvent for $\frac{\partial u_h}{\partial n}$.

In the two-dimensional case N = 2, such sets are for example

$$\frac{\partial u_h}{\partial n} \in P_2(K') \qquad (18)$$

$$\frac{\partial u_h}{\partial n} \in P_3(K') \qquad (19)$$

For notation of degrees of freedom, (.), (|) and (//) symbolize the values of the function, the derivative in normal direction and the mixed normal-tangential derivative, respectively.

A simple V_h may be constructed from Hermite-P_3-element with 10 degrees of freedom.

$u_h \in P_3$ (K)

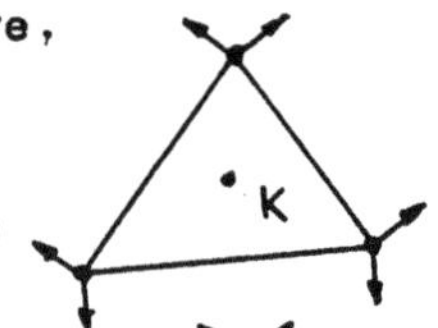

Here (reg 1) is fulfilled if in addition the normal derivative is continuous in the middle of K'. (cf. (18)). We get overdeterminated finite element trial functions defined by one additional linear restriction for each K'.

$u_h \in P_3$ (K)

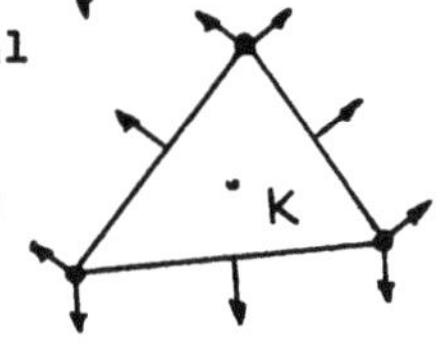

If Ω is a convex polynomial with m edges which is triangulated into n triangles, we get dim V_h = n-m+3. On this account we were able to fulfill the inequalities (15) elementwise only for triangular Ω. For that reason, Werner [8] gives an alternate method. He takes the inequalities (15) and (16) (in the nodes) for side restriction in an optimization process and minimizes

$$\bar{u}_h(x) - \underline{u}_h(x) \longrightarrow \min \qquad (20)$$

in a fixed point $x \in \Omega$.

4. The Bogner-Fox-Schmit rectangle

In the one-dimensional case the cubic C^1-element can be chosen to construct enclosures of the solution of (1). In that case, the choice

P_3 (K)

case N = 1

$$F_h = \tilde{F}_h = \{ f_h \mid f_K \in P_k(K) \}$$

is possible at which k = 0 or k = 1. Our aim is to construct analogous elements in the multidimensional case.

A "Cartesian product" of such one-dimensional cubic C^1-elements yields an element of 4^N degrees of freedom where every 2^N

degrees of freedom are located in each corner of the element. (reg 2) is fulfilled. In two dimensions the element constructed in this way is the Bogner-Fox-Schmit rectangle [2].

Q_3 (K)

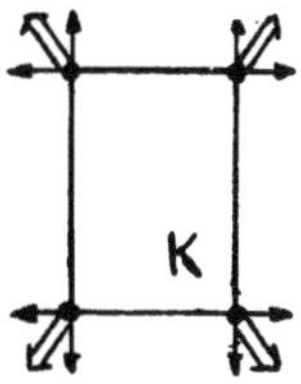

Case N = 2

For realization of (13), we discuss the set

$$F_K = -\Delta Q_3\ (K). \tag{21}$$

Let every K be described by four quantities x_o, y_o, h_x, h_y with

$$K = \{(x,y) \mid |x-x_o| \le h_x, |y-y_o| \le h_y\}$$

$$h_K = h_x + h_y$$

and set

$$\begin{aligned} \xi &= \xi(x,y) = x+y-x_o-y_o \in I_\xi = [-h_K, h_K] \\ \eta &= \eta(x,y) = x-y-x_o-y_o \in I_\eta = [-h_K, h_K] \\ \zeta &= \zeta(x,y) = \xi^4-\eta^4 \quad \in I_\zeta = [-h_K^4, h_K^4] \end{aligned} \tag{22}$$

By checking (21) we get the following result.

<u>Lemma 1:</u> $F_K = P_3\ (I_\xi) + P_3\ (I_\eta) + P_1\ (I_\zeta)$, i. e.

$$Q_1\ (K) \subset F_K \subset P_4\ (K), \tag{23}$$

and the dimension of F_K is 8∎

Therefore, we choose for k = 1 or k = 2

$$\tilde{F}_h = \{f_h \mid f_K \in Q_{k-1}\ (K)\} \tag{24}$$

and obtain two different algorithms for k = 1,2,respectively.

Let

$$\begin{aligned} -\Delta \underline{u}_K &= \underline{f}_K = \underline{a} + \underline{b}_1\xi + \underline{b}_2\xi^2 + \underline{b}_3\xi^3 + \underline{c}_1\eta + \underline{c}_2\eta^2 + \underline{c}_3\eta^3 + \underline{d}\zeta \\ -\Delta \bar{u}_K &= \bar{f}_K = \bar{a} + \bar{b}_1\xi + \bar{b}_2\xi^2 + \bar{b}_3\xi^3 + \bar{c}_1\eta + \bar{c}_2\eta^2 + \bar{c}_3\eta^3 + \bar{d}\zeta \end{aligned} \tag{25}$$

$$z^+ = \max\{z, 0\} \qquad z^- = \min\{z, 0\} \tag{26}$$

<u>Algorithm 1</u>

(1a) Determine a piecewise constant enclosure of f such that

$$\underset{\sim}{f}_h \le f \le \tilde{f}_h \qquad \underset{\sim}{f}_K,\quad \tilde{f}_K \in Q_o\ (K).$$

Let $\underset{\sim}{f}_K = \underset{\sim}{a}$, $\tilde{f}_K = \tilde{a}$.

(1b) Solve the systems arising from the equations

$$\underline{b}_3 = \underline{c}_3 = \underline{d} = 0 \qquad \bar{b}_3 = \bar{c}_3 = \bar{d} = 0$$

$$\underline{a} = \underset{\sim}{a} - (\underline{b}_1^+ + \underline{c}_1^+)h_k - (\underline{b}_2^+ + \underline{c}_2^+)h_k^2 \qquad \bar{a} = a - (\bar{b}_1^- + \bar{c}_1^-)h_k - (\bar{b}_2^- + \bar{c}_2^-)h_k^2$$

on each element.

Remark 1: The first three linear equations of (1b) restrict the trial space to $Q_3(K) \cap P_4(K)$.

Algorithm 2

(2a) Determine a piecewise bilinear (noncontinuous) enclosure of f such that

$$\underset{\sim}{f}_h \le f \le \tilde{f}_h \qquad \underset{\sim}{f}_K, \quad \tilde{f}_K \in Q_1(K)$$

Let

$$\underset{\sim}{f}_K = \underset{\sim}{a} + \underset{\sim}{b}\xi + \underset{\sim}{c}\eta + \underset{\sim}{d}(\xi^2-\eta^2), \qquad \tilde{f}_K = \tilde{a} + \tilde{b}\xi + \tilde{c}\eta + \tilde{d}(\xi^2-\eta^2).$$

(2b) Solve the systems arising form the equations

$$\underline{b}_1 = \underset{\sim}{b} \qquad \bar{b}_1 = \tilde{b}$$

$$\underline{c}_1 = \underset{\sim}{c} \qquad \bar{c}_1 = \tilde{c}$$

$$\underline{b}_2 - \underline{c}_2 = 2\underset{\sim}{d} \qquad \bar{b}_2 - \bar{c}_2 = 2\tilde{d}$$

$$\underline{a} = \underset{\sim}{a} - (\underline{b}_2 + \underline{c}_2)^+ h_K^2 - (\underline{b}_3^+ + \underline{c}_3^+)h_K^3 - \underline{d}^+ h_K^4 \qquad \bar{a} = \tilde{a} - (\bar{b}_2 + \bar{c}_2)^- h_K^2 - (\bar{b}_3^- + \bar{c}_3^-)h_K^3 - \bar{d}^- h_K^4$$

From construction it follows $\underline{f}_h \le f \le \bar{f}_h$. Thus, we get an enclosure

$$\underline{u}_h \le u \le \bar{u}_h. \tag{27}$$

Now, we can formulate the following error estimation.

Theorem 2: We suppose $\underline{u}_h$, $\bar{u}_h$ to be a solution of the algorithm k (k = 1,2) and the enclosure of the right hand side in the first step to be optimal. Let $f \in W^{k,p}(\Omega)$, $k < p \le \infty$. Then it holds

$$\|\bar{u}_h - \underline{u}_h\|_{0,\infty} \le C h^{k-\frac{2}{p}} \tag{28}$$

if $h = \max_{K \in T_h} h_K$ ∎

Proof: For piecewise $Q_{k-1}(K)$ interpolation f_h, we get the estimate

$$\|f - f_h\|_{0,\infty,K} \le C_1 h^{k-\frac{2}{p}} |f|_{k,p,K} \le C_1 h^{k-\frac{2}{p}} |f|_{k,p,\Omega} =: \eta$$

([3] , Theorem 3.1.5.). Therefore,

$$f_h - \eta \le \underset{\sim}{f}_h \le f \le \tilde{f}_h \le f_h + \eta .$$

From construction of $\underline{f}_h$, $\bar{f}_h$ the estimates

$$\|\underline{f}_h - \tilde{f}_h\|_{0,\infty,K} \le c_2 h^k \qquad \|\bar{f}_h - \tilde{f}_h\|_{0,\infty,K} \le c_2 h^k$$

follow. Let $s \in C^2(\Omega) \cap C(\bar{\Omega})$ be a barrier-function which suffices $s \ge 1$, $-\Delta s \ge 1$. Thus, it holds

$$|-\Delta(\bar{u}_h - \underline{u}_h)| = |\bar{f}_h - \underline{f}_h| \le 2(c_2 h^k + \eta) \le c_3 h^{k-\frac{2}{p}}(-\Delta s)$$

and (28) follows from Theorem 1, where $c = c_3 \; \|s\|_{0,\infty}$.

In the proof we use the -regularity of f only on elements. Hence, Theorem 2 is true as well in the case of noncontinuous f, if the noncontinuity is located only on the net of triangulation.

For a simple numerical example we consider the unique square $\Omega = [0,1]^2$ and a triangulation into nxn subsquares. We apply algorithm 1 to the Dirichlet problem

$$\begin{aligned} -\Delta u &= 1, && \text{in } \Omega, \\ u &= 0 && \text{on } \partial\Omega. \end{aligned} \tag{29}$$

The numerical results seem to sustain linear convergence.

n	1	2	4	8
$\|\bar{u}_h - \underline{u}_h\|_{0,\infty}$	.063	.031	.019	.010

5. Concluding remarks

Practicable algorithms for computing sub- and super-solutions for the Dirichlet problem

$$\begin{aligned} -\Delta u &= f && \text{in } \Omega \\ u &= g && \text{on } \partial\Omega \end{aligned} \tag{30}$$

have been formulated. The algorithms are suitable for combination with the monotone iteration discretization idea [1]. On this base methods for solving (1) should be constructed.

Another approach would be a combination with the method of defect shifting (cf. [7]) where we can use a finite element supersolution of (29) to construct effectively enclosures for (30).

[1] Abhari, H., Al-Zanaidi, M., Großmann, Ch.: Computational Aspects of the Monotone Iteration Discretization Algorithm, Preprint 07-01-88, TU Dresden, 1988

[2] Bogner, F., K., Fox, R. L., Schmit, L. A.: The generation of interelement compatible stiffness and mass matrices by the use of interpolation formulas, Proc. Conf. Matrix Methods in Structural Mechanics, Ohio 1965

[3] Ciarlet, Ph.: The Finite Element Method for Elliptic Problems, Amsterdam, 1978

[4] Glashoff, K., Werner, B.: Inverse Monotonicity of Monotone L-operators with Application to Quasilinear and Free Boundary Problems, J. Anal. Appl. 72 (1979) 1, 89 - 105

[5] Großmann, Ch.: Monotone Discretization of two-Point Boundary Value Problems and Related Numerical Methods. In: Adams et. al. (eds): Discretization in Differential Equations and Enclosures, Berlin 1987

[6] Ladyzenskaya, O. A., Ural'ceva N. N.: Linejnye i Kvazilinejnye uravnenija ellipticeskogo tipa, Moskva 1973

[7] Roos, H.-G., this proceedings

[8] Werner, B.: Monotonie und finite Elemente bei elliptischen Differentialgleichungen, ISNM 27, 1975, 309 - 329

COMPOSITE GEOMETRIC CONTINUOUS LAGRANGE-HERMITE CURVES

Gerhard Geise [1]

0. Introduction. Provissionally let a curve element be a point or a point together with a smooth and oriented curve passing through the point possessing some (differential-)geometric properties at the point; the point is called the support of the curve element. It is a typical problem in geometric design to join a sequence of curve elements by curves fulfilling the given geometric properties at the supports. Also one speaks on designing curves passing through the curve elements. Often one demands analytic or geometric continuity of some class r (C^r, G^r).

Nowadays a curve outlined just will be constructed as a composite curve, the curve sections or spans joining consecutive pairs of curve elements. There are most powerful techniques. The so-called cubic spline interpolation is well known to join a sequence of points by C^2-curves. This method is applied in curve design too, especially in a modified manner securing independence on the choice of co-ordinate system. Furthermore, a sequence of curve elements of 1-st order or tangent elements (each constituted by a point and an associated direction unit vector) is usually joined by a sequence of Bézier curves guaranteeing G^1- or G^2-continuity (known as generalized parametric Hermite interpolation too; Bär 1977).

Entire rational functions used by the mentioned techniques will remain in our field of views. In practice there are problems of geometric curve design demanding the construction of composite curves by spans joining more than two curve elements. Our main aim is to consider the concept of curve elements (section 1). To join an ordered set of curve elements by a single curve here we apply the method of linear combination by polynomial blending functions. The set-up (section 2) leads directly to the well known Lagrange-Hermite technique described in Werner/Schaback 1971, for instance; therefore we do not execute it here. Although the Lagrange-Hermite method has a bad reputation the progress made both in arithmetic of computers and in methods of numerical analysis allow to test it newly. The achieved results are rather promising. There are experiences made in common with other authors, and we shall report occasionally. Other techniques for instance Bézier are also used but comparisons had not been made.

1. Curve elements. A curve element (briefly element) of order 0 is a single point p_0. An element of order $r > 0$ with the support p_0 is a class of curves where each curve (called curve element of order r too) is given by a parametric representation of the kind

[1] Technische Universität Dresden, Mommsenstr. 13, DDR-8027 Dresden

$$x(u) = p_o + \sum_{i=1}^{r} \frac{u^i}{i!} \cdot p_i \quad , \tag{1}$$

defined on a neighbourhood of zero, and determining an algebraic curve of order r. Here we assume that for $r = 1$, $r = 2$, $r = 3$ holds: $p_1 \neq o$, p_1 and p_2 respectively p_1, p_2 and p_3 are linearly independent. Any two curves of the class, say (1) and

$$y(v) = p_o + \sum_{i=1}^{r} \frac{v^i}{i!} \cdot q_i \quad , \tag{2}$$

are connected by a regular transformation of the derivatives at p_o, and reads as follows:

$$\begin{bmatrix} q_1 \\ q_2 \\ \cdot \\ \vdots \\ q_r \end{bmatrix} = \begin{bmatrix} a_1 & 0 & \cdot & \cdot & \cdot & 0 \\ a_{r+1} & a_2 & \cdot & & & \vdots \\ \vdots & & \ddots & & & \cdot \\ & & & \ddots & & 0 \\ a_R & \cdot & \cdot & \cdot & a_{2r-1} & a_r \end{bmatrix} \begin{bmatrix} p_1 \\ p_2 \\ \cdot \\ \vdots \\ p_r \end{bmatrix} \quad , \tag{3}$$

where first of all $R = r(r+1)/2$ and $a_1 \cdot a_2 \cdot \ldots \cdot a_r \neq 0$. The transformations (3) form a group depending on p parameters, $p \leq R$, preserving the considered geometric properties. Hence we describe a curve element as a class of curves by means of one curve of the class, and the curve elements of fixed order r are decomposed in subclasses. A classification of curve elements of order r = 1, 2, 3, which are essential in applications is given in a hierarchic oder (* means: no condition; lines in brackets are to refere to historical notions):

$$\begin{array}{l} r = 1 \\ r = 2 \\ r = 3 \end{array} \quad \begin{bmatrix} a_1 & 0 & 0 \\ a_3 & a_2 & 0 \\ a_6 & a_5 & a_4 \end{bmatrix}, \quad a_1 \cdot a_2 \cdot a_4 \neq 0:$$

r	a_1	a_2	a_3	a_4	a_5	a_6	invariance of
1	$\neq 0$						tangent (at p_o)
	>0						orientation of the curve: G^1
	(1						C^1)
2	>0	$\neq 0$	*				osculating plane
	>0	>0	*				osculating half plane, containing centre of curvature
	>0	a_1^2	*				centre of curvature: G^2
	(>0	a_1^2	0				visual continuous)
	(1	1	0				C^2)

r	a_1	a_2	a_3	a_4	a_5	a_6	invariance of
3	>0	≠0	*	a_1a_2	*	*	torsion, but not curvature
	>0	a_1^2	*	a_1^3	*	*	curvature and torsion
	>0	a_1^2	*	a_1^3	$3a_1a_2$	*	osculating sphere: G^3
	(>0	a_1^2	0	a_1^3	0	0	visual continuous)
	(1	1	0	1	0	0	C^3)

The strongest geometric property is G^r and means contact of order r. This is mostly used. The other curve elements of order r are roughly summarized as curve elements 'of fractional order' r. The strongest analytic property is C^r.

We consider the case G^r only. There are r free parameters. Additionally, the parameter associated with the support is free. Therefore one has r+1 free parameters at each G^r-curve element named shape or intrinsic parameters.

2. Joining of curve elements by Lagrange-Hermite curves. If one has k_r curve elements of order r and r runs from 0 to R than there are to join

$$M = k_o + k_1 + \dots + k_R$$

curve elements depending on

$$N+1 = k_o + 2\cdot k_1 + \dots + (R+1)\cdot k_R$$

free parameters. The set-up for joining the given curve elements using polynomial blending functions $f_{ij}^r(t)$ $(j = 0,\dots,r;\ i \in I_r,\ I_r$ suitable index sets; $r = 0,\dots,R)$ is

$$x(t) = \sum_{r=0}^{R} \sum_{i \in I_r} \sum_{j=0}^{r} f_{ij}^r(t)\cdot q_{ij}^r \ ,$$

where the q_{ij}^r $(j = 1,\dots,r)$ characterize the i-th curve element of order r with the support q_{i0}^r. Using the Lagrange-Hermite technique we obtain a solution immediately: $x(t) = x(t;t_1,\dots,t_M)$. Substituting the $q_{i1}^r, \dots, q_{ir}^r$ by the p_{ij}^r according to (3) for each i corresponding to a curve element of order $r > 0$ the full manifold of curves through the given curve elements results.

Clearly a large number of free parameters is unfavourable. To solve problems of approximation or optimization outstanding now, statements and methods of numerical analysis are needed. Free-form curves (and surfaces especially) are of interest only so far as the shape parameters are suitable to overcome the problems.

Bär, G.: Parametrische Interpolation empirischer Raumkurven. Z. Angew. Math. Mech. (ZAMM) 53(1977) 433-440.

Werner, H.; Schaback, R.: Praktische Mathematik II. Springer-Verlag Berlin Göttingen Heidelberg 1972.

2D-INTERPOLATION BY BIQUADRATIC SPLINES WITH MINIMAL SURFACE

Anke Hänler and Gerhard Maess [1]

The aim of this paper is the construction of smooth interpolating biquadratic splines minimizing certain functionals one of which may be interpreted as a measure of the surface. Related functionals were used by other authors for splines of higher degree (cf. for instance /1/, /5/). In our case all the parameters of the spline are computed recursively or by explicit formulae. There is no need for solving systems of linear equations (as e.g. in /4/).

1. Introduction

Let $\mathbb{D}$ denote a rectangular domain in the (x,y)-plane

$$\mathbb{D} := [a\ ,\ b] \times [c\ ,\ d] \tag{1}$$

with the partition

$$\begin{aligned} \pi_{\mathbb{D}} : \quad & a = x_0 < x_1 < \ldots < x_M = b, \quad c \leq y \leq d, \\ & c = y_0 < y_1 < \ldots < y_N = d, \quad a \leq x \leq b. \end{aligned} \tag{2}$$

Suppose there are given the values

$$f_{ij} := f(x_i,y_j), \quad i=0(1)M, \quad j=0(1)N, \tag{3}$$

of a real-valued function f defined on $\mathbb{D}$. The biquadratic spline p, $p \in C^1(\mathbb{D})$, we are looking for, then may be written in the form

$$\begin{aligned} p(x,y) = p(x_{i-1}+sh_i, y_{j-1}+tk_j) = w(t)^T A_{ij}\, w(s), \quad 0 \leq s,t \leq 1\ , \\ x_{i-1} \leq x \leq x_i\ , \quad y_{j-1} \leq y \leq y_j\ , \quad i=0(1)M, \quad j=0(1)N, \end{aligned} \tag{4}$$

where w denotes the vector of the local quadratic basic functions,

$$w(t) := (w_1(t),\ w_2(t),\ w_3(t))^T = (1-t^2,\ t^2,\ t(1-t))^T,\ 0 \leq t \leq 1, \tag{5}$$

and h_i and k_j are the step-sizes in the x- and y-direction, respectively

$$h_i := x_i - x_{i-1}\ , \quad k_j := y_j - y_{j-1}\ , \quad i=1(1)M, \quad j=1(1)N. \tag{6}$$

With the abbreviations

$$b_{ij} := p_x(x_i,y_j), \quad d_{ij} := p_y(x_i,y_j), \quad e_{ij} := p_{xy}(x_i,y_j) \tag{7}$$

the matrices A_{ij} take the form

$$A_{ij} := \begin{pmatrix} f_{i-1,j-1} & f_{i,j-1} & h_i b_{i-1,j-1} \\ f_{i-1,j} & f_{i,j} & h_i b_{i-1,j} \\ k_j d_{i-1,j-1} & k_j d_{i,j-1} & h_i k_j e_{i-1,j-1} \end{pmatrix} . \tag{8}$$

---

[1] Wilhelm-Pieck-Universität, Sektion Mathematik, Universitätsplatz 1, Rostock, DDR - 2500

The continuity conditions for the first derivatives

$$p_x(x_i-0,y_j) = p_x(x_i+0,y_j), \quad p_y(x_i,y_j-0) = p_y(x_i,y_j+0),$$
$$p_{xy}(x_i,y_j-0) = p_{xy}(x_i,y_j+0), \; p_{xy}(x_i-0,y_j) = p_{xy}(x_i+0,y_j) \tag{9}$$

yield the following recurrences for the unknown parameters b_{ij} , d_{ij} , and e_{ij}

$$b_{ij} = 2g_{xij} - b_{i-1,j}\,, \quad g_{xij} := (f_{ij} - f_{i-1,j})/h_i, \; i=1(1)M, \; j=0(1)N,$$

$$d_{ij} = 2g_{yij} - d_{i,j-1}\,, \quad g_{yij} := (f_{ij} - f_{i,j-1})/k_j, \; j=1(1)N, \; i=0(1)M,$$

$$e_{ij} = 2g_{xyij} - e_{i,j-1}\,, \quad g_{xyij} := (b_{ij} - b_{i,j-1})/k_j, \; j=1(1)N, \; i=0(1)M,$$

$$e_{ij} = 2g_{yxij} - e_{i-1,j}\,, \quad g_{yxij} := (d_{ij} - d_{i-1,j})/h_i, \; i=1(1)M, \; j=0(1)N.$$

It can be proved (cf. /3/, pp. 134 - 135) that all parameters are determined uniquely, if there are given the following initial values

$$b_{oj}\,, \; j=0(1)N, \quad d_{io}\,, \; i=0(1)M, \quad e_{oo}\,. \tag{10}$$

As in /2/, /3/ we choose the parameters b_{oj} and d_{io} such that p minimizes the functionals

$$Q_{xj}(p) := \int_a^b K^2(p(x,y_j))dx \quad \text{and} \quad Q_{yi}(p) := \int_a^b K^2(p(x_i,y))dy \tag{11}$$

over the gridlines $y=y_j$, $j=0(1)N$, or $x=x_i$, $i=0(1)M$, respectively. Here $K(f(x))$ denotes the (approximated) curvature of $z=f(x)$ at the point x:

$$K(p(x,y_j)) := (1 + g_{xij}^2)^{-3/2} \, p_{xx}(x,y_j), \quad a \le x \le b, \tag{12}$$

$$K(p(x_i,y)) := (1 + g_{yij}^2)^{-3/2} \, p_{yy}(x_i,y), \quad c \le y \le d. \tag{13}$$

So the minimization of (11) assures a minimal total curvature of the spline p along the horizontal ($y=y_j$, $j=0(1)N$) and the vertical ($x=x_i$, $i=0(1)M$) gridlines. For fixing the remaining parameter e_{oo} we used in /2/, /3/ one of the functionals

$$Q_x(p) := \int_a^b K^2(p_y(x,y_l))dx \quad \text{or} \quad Q_y(p) := \int_a^b K^2(p_x(x_l,y))dy\,, \tag{14}$$

where l denotes an arbitrary chosen integer between 0 and N or M, resp. The aim is that one of the first partial derivatives p_y or p_x has a minimal total curvature along an arbitrary fixed horizontal (or vertical) gridline. In this case the determination of e_{oo} is very easy, but the result depends on the choice of l. To overcome this drawback, we now use more general functionals of the type

$$Q(p) = \iint_D \sum_{0\le\alpha,\beta\le 2} c_{\alpha\beta} \left(\frac{\partial^{\alpha+\beta} p(x,y)}{\partial x^\alpha \partial y^\beta}\right)^2 dxdy, \quad c_{\alpha\beta} = 0\,. \tag{15}$$

In the special case $c_{10} = c_{01} = 1$ (and $c_{\alpha\beta} = 0$ else) we get the functional

$$Q_1(p) = \iint_D (p_x^2 + p_y^2)\, dx\, dy \tag{16}$$

which may be interpreted as a measure of the surface of

$$z = p(x,y), \quad (x,y) \in \mathbb{D}. \tag{17}$$

Other interesting special cases are

$$Q_2(p) := \iint_{\mathbb{D}} p_{xxyy}^2 \, dx \, dy \tag{18}$$

or the "thin-plate-functional" (cf. /5/)

$$Q_3(p) := \iint_{\mathbb{D}} (p_{xx}^2 + 2 p_{xy}^2 + p_{yy}^2) \, dx \, dy . \tag{19}$$

2. Minimization of the functional Q(p)

In the following sections we will better write the spline p (over each subrectangle $[x_{i-1}, x_i] \times [y_{j-1}, y_j]$) in the form

$$p(x,y) = w(t)^T A_{ij}^o w(s) + h_i k_j e_{i-1,j-1} w_3(t) w_3(s), \quad 0 \leq s,t \leq 1, \tag{20}$$

with

$$A_{ij}^o := \begin{pmatrix} f_{i-1,j} & f_{i,j-1} & h_i b_{i-1,j-1} \\ f_{i-1,j} & f_{ij} & h_i b_{i-1,j} \\ k_j d_{i-1,j-1} & k_j d_{i,j-1} & 0 \end{pmatrix} .$$

So we obtain for the derivatives

$$\frac{\partial^{\alpha+\beta} p(x,y)}{\partial x^\alpha \partial y^\beta} = h_i^{-\alpha} k_j^{-\beta} (w^{(\beta)}(t)^T A_{ij}^o w^{(\alpha)}(s) + h_i k_j e_{i-1,j-1} w_3^{(\beta)}(t) \cdot w_3^{(\alpha)}(s)).$$

By summation over all α,β and over all subrectangles and by integration we get (cf. (15))

$$Q(p) = \sum_{\alpha,\beta} c_{\alpha\beta} \sum_{i=1}^{M} \sum_{j=1}^{N} h_i^{1-2\alpha} k_j^{1-2\beta} \cdot \int_0^1 \int_0^1 (w^{(\beta)}(t)^T A_{ij}^o w^{(\alpha)}(s) + h_i k_j e_{i-1,j-1} w_3^{(\beta)}(t) w_3^{(\alpha)}(s))^2 \, ds dt. \tag{21}$$

From the e_{ij}-recurrences (cf.(9)) it follows that all the e_{ij} depend linearly on e_{oo}, more precisely it holds

$$e_{ij} = m_{ij} + (-1)^{i+j} e_{oo} \tag{22}$$

with

$$m_{ij} = \begin{cases} 0 & \text{for } i=j=0 , \\ 2 \sum_{l=1}^{i} (-1)^{i+l} g_{yxlo} & \text{for } j=0 \text{ and } i>0 , \\ 2 \sum_{l=1}^{j} (-1)^{j+l} g_{xyol} & \text{for } i=0 \text{ and } j>0 , \\ 2 \sum_{l=1}^{i} (-1)^{i+l} g_{yxlj} + 2 \sum_{l=1}^{j} (-1)^{i+l+j} g_{xyol} & \text{for } i,j > 0 . \end{cases} \tag{23}$$

So we get the minimizer of the functional (21) from the vanishing first derivative of $Q(p)$ (as a function of e_{oo}),

$$\frac{dQ(p)}{de_{oo}} = 2\sum_{\alpha,\beta} c_{\alpha\beta} \sum_{i=1}^{M} \sum_{j=1}^{N} h_i^{1-2\alpha} k_j^{1-2\beta} \cdot$$

$$\cdot \int_0^1\int_0^1 (w^{(\beta)}(t)^T A_{ij}^o w^{(\alpha)}(s) + h_i k_j e_{i-1,j-1} w_3^{(\beta)}(t) w_3^{(\alpha)}(s)) \cdot \tag{24}$$

$$\cdot (-1)^{i+j} h_i k_j w_3^{(\beta)}(t) w_3^{(\alpha)}(s) ds dt = 0,$$

since the second derivative is greater than zero (for $c_{\alpha\beta} > 0$):

$$\frac{d^2Q(p)}{de_{oo}^2} = 2\sum_{\alpha,\beta} c_{\alpha\beta} \sum_{i=1}^{M}\sum_{j=1}^{N} h_i^{3-2\alpha} k_j^{3-2\beta} \int_0^1 w_3^{(\alpha)^2}(s) ds \int_0^1 w_3^{(\beta)^2}(t) dt > 0.$$

Using (22) and the abbreviations

$$I_{ij}^{\alpha\beta} := \left(\int_0^1 w_3^{(\beta)}(t) w^{(\beta)}(t) dt\right)^T A_{ij}^o \left(\int_0^1 w_3^{(\alpha)}(s) w^{(\alpha)}(s) ds\right) \tag{25}$$

and

$$J^{\alpha\beta} := \int_0^1 w_3^{(\alpha)^2}(s)\, ds \cdot \int_0^1 w_3^{(\beta)^2}(t)\, dt \tag{26}$$

we obtain from (24) the following explicit representation for the optimal parameter e_{oo}

$$e_{oo} = \frac{\sum_{0\le\alpha,\beta\le 2} c_{\alpha\beta} \sum_{i=1}^{M}\sum_{j=1}^{N} (-1)^{i+j+1} h_i^{2-2\alpha} k_j^{2-2\beta} (I_{ij}^{\alpha\beta} + h_i k_j m_{ij} J^{\alpha\beta})}{\sum_{0\le\alpha,\beta\le 2} c_{\alpha\beta} J^{\alpha\beta} \sum_{i=1}^{M}\sum_{j=1}^{N} h_i^{3-2\alpha} k_j^{3-2\beta}}. \tag{27}$$

3. The minimal-surface-functional $Q_1(p)$

In the case of $Q_1(p)$ we have

$$c_{\alpha\beta} = \begin{cases} 1 & \text{for } (\alpha=1 \text{ and } \beta=o) \text{ or } (\alpha=o \text{ and } \beta=1), \\ 0 & \text{else.} \end{cases}$$

So we need the integrals (cf. (25) and (5))

$$\int_0^1 (w_3(s))^2 ds = \frac{1}{30}, \qquad \int_0^1 (w_3'(s))^2 ds = \frac{1}{3},$$

$$\int_0^1 w_3(s) w(s)^T ds = \frac{1}{60}(7, 3, 2) =: \frac{1}{60} a^T, \tag{28}$$

$$\int_0^1 w_3'(s) w'(s)^T ds = \frac{1}{3}(1, -1, 1) =: \frac{1}{3} v^T,$$

to obtain

$$I_{ij}^{01} = v^T A_{ij}^o u / 180 , \quad I_{ij}^{10} = u^T A_{ij}^o v / 180, \quad J^{01} = J^{10} = 1/90 ,$$

and the optimal parameter e_{oo} takes the form

$$e_{oo} = \frac{\sum_{i=1}^{M} \sum_{j=1}^{N} (-1)^{i+j+1}(k_j^2 u^T A_{ij}^o v + h_i^2 v^T A_{ij}^o u + 2m_{ij}h_ik_j(h_i^2+k_j^2))}{2 \sum_{i=1}^{M} \sum_{j=1}^{N} h_i k_j (h_i^2 + k_j^2)} . \tag{29}$$

4. Minimization of the functional $Q_2(p)$

In the case of $Q_2(p)$ we have

$$c_{\alpha\beta} = \begin{cases} 1 & \text{for } \alpha = \beta = 2 , \\ 0 & \text{else} . \end{cases}$$

For the integrals in (25) we now obtain

$$\int_0^1 w_3''(s) w''(s) ds = 4 v, \quad v := (1, -1, 1)^T . \tag{30}$$

So it follows

$$I_{ij}^{2,2} = 16 v^T A_{ij}^o v, \quad J^{2,2} = 16,$$

and the optimal parameter e_{oo} now reads

$$e_{oo} = \frac{\sum_{i=1}^{M} \sum_{j=1}^{N} (-1)^{i+j+1} (h_i^{-2} k_j^{-2} v^T A_{ij}^o v + h_i^{-1} k_j^{-1} m_{ij})}{\sum_{i=1}^{M} \sum_{j=1}^{N} h_i^{-1} k_j^{-1}} . \tag{31}$$

5. The thin-plate-functional $Q_3(p)$

In the case of $Q_3(p)$ we have

$$c_{o\beta} = \begin{cases} 1 & \text{for } (\alpha = 2 \text{ and } \beta = 0) \text{ or } (\alpha = 0 \text{ and } \beta = 2) , \\ 2 & \text{for } \alpha = \beta = 1 , \\ 0 & \text{else}, \end{cases}$$

and from (28) and (30) it follows

$$I_{ij}^{2,0} = u^T A_{ij}^o v / 15, \quad I_{ij}^{0,2} = v^T A_{ij}^o u / 15, \quad I_{ij}^{1,1} = v^T A_{ij}^o v / 9,$$

$$J^{2,0} = J^{0,2} = 2/15, \quad J^{1,1} = 1/9.$$

So the optimal parameter e_{oo} takes the form

$$w_{ij} := 3 h_i^{-1} k_j^3 + 5 h_i k_j + 3 h_i^3 k_j^{-1} ,$$

$$e_{oo} = \frac{\sum_{i=1}^{M} \sum_{j=1}^{N} (-1)^{i+j+1}(3h_i^{-2}k_j^2 u^T A_{ij}^o v + 5v^T A_{ij}^o v + 3h_i^2 k_j^{-2} v^T A_{ij}^o u + w_{ij} m_{ij})}{\sum_{i=1}^{M} \sum_{j=1}^{N} w_{ij}} .$$

Error estimations (for the case that the interpolated function f is sufficiently smooth) are left to a forthcoming paper.

6. Examples

For the computation of the surface in Fig. 1 we used the domain $\mathbb{D} = [-1, 1] \times [-1, 1]$ with a non-equidistant 3×3-partition, given together with the values f_{ij} in the following table

y_j	-1.0	-0.6	0.6	1.0	x_i
1.0	0.0	0.4	0.4	0.0	3
0.0	0.5	.64	.64	0.5	2
-0.6	0.4	.55	.55	0.4	1
-1.0	0.0	0.4	0.4	0.0	0
i	0	1	2	3	j

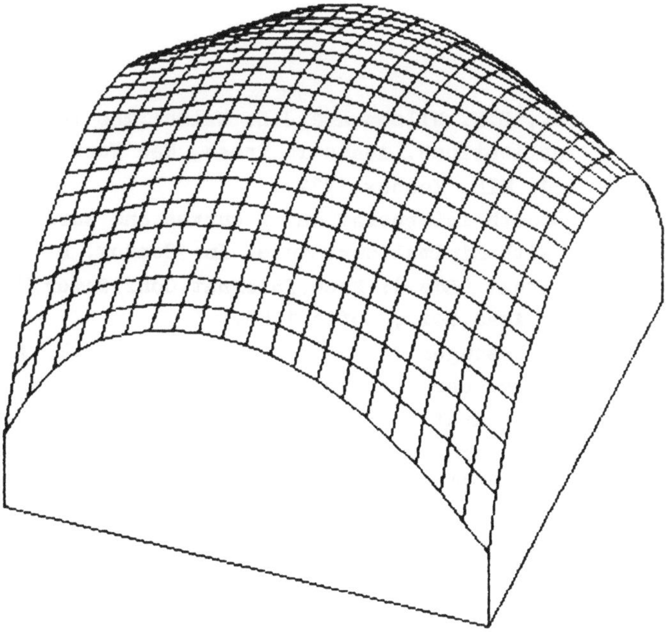

Fig. 1

For the surface in Fig. 2 we used the same domain but an equidistant 3×3-partition and random values f_{ij} , $i,j=0(1)3$. In both examples the optimal parameter e_{00} was determined by (23), i.e. by minimization of the "surface-functional" $Q_1(p)$.

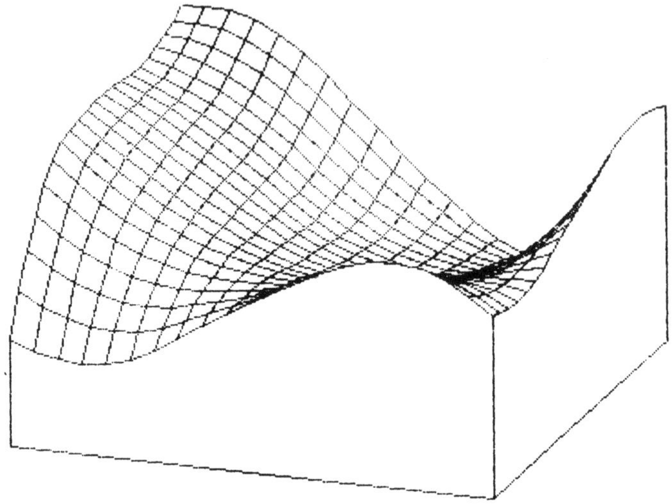

Fig. 2

7. References

/1/ Ahlberg, J.H., E.N. Nilson and J.L. Walsh: The Theory of Splines and their Applications. Academic Press, New York 1967 (Russ. translation: Izd. MIR, Moscow 1972).

/2/ Maess, G.: Smooth interpolation of curves and surfaces by quadratic splines with minimal curvature. In: Sendov, B., R. Lazarov and P. Vasilevski (Eds.): Proc.Internat.Conf.Numer.Math.Appl.,pp. 75-81, Bulgar.Akad.Sci., Sofia 1985.

/3/ Maess, G.: Vorlesungen über numerische Mathematik. Akademie-Verl. Berlin und Birkhäuser Verlag Basel 1988.

/4/ Mettke, H.: Biquadratische Interpolations- und Volumenabgleichssplines. Beiträge zur Numer. Math. 9(1981),135-145.

/5/ Schumaker, L.L.: Fitting surfaces to scattered data. In: Lorentz, G.G., C.K.Chui and L.L.Schumaker (Eds.): Approximation Theory II, pp. 203-269, Academic Press, New York 1976.

SPLINES UND DIE WÄRMELEITUNGSGLEICHUNG

von

Günter Meinardus[1)]

Seit langem ist bekannt, daß Räume polynomialer Splinefunktionen in einer reellen Variablen auf gewisse Haar'sche Räume bijektiv abgebildet werden können (vgl. [5]). Die zur Abbildung verwendeten Integraltransformationen erlauben es, über das Vorzeichenverhalten von Splines, insbesondere bei Aufgaben der gleichmäßigen Approximation, Aussagen zu beweisen. Diese Untersuchungen blieben bisher jedoch vollständig im theoretischen Bereich. Es stellt sich die Frage, ob diese Transformationen auch dann Vorteile bieten, wenn konkrete Fragen behandelt werden sollen, z.B. wenn Konstruktionsverfahren zur Gewinnung bester Approximationen gesucht werden, oder wenn man an a priori Abschätzungen der Minimalabweichung interessiert ist.

In der vorliegenden Note wird über Ergebnisse zu diesem Komplex berichtet. Es ergeben sich gleichzeitig konstruktive Verfahren zur Lösung von Anfangswertaufgaben der Wärmeleitungsgleichung in einer einzigen Raumvariablen. Hier treten durch Verwendung einfacher Rekursionen, abgesehen vom Approximationsfehler der Anfangsfunktion, prinzipiell keine Verfahrensfehler auf. Ferner liegen einige Überlegungen zum Approximationsproblem mit Splines bei sog. freien Knoten nahe (vgl. [6]). Es geht hier i.w. um Aussagen über den Tangentialraum der betrachteten parametrisierten Funktionenmenge. – Für spezielle Funktionen kann man aus einem neuen Vergleichssatz Einschließungen numerischer und asymptotischer Art für die Minimalabweichung erhalten.

Bei allen diesen Untersuchungen beschränken wir uns auf spezielle Integraltransformationen, auf die klassischen polynomialen Splines und auf die gewöhnliche Tschebyscheff-Approximation. Die meisten Resultate können offensichtlich verallgemeinert werden. Eine auf möglichst große Allgemeinheit Wert legende Darstellung wurde bewußt vermieden, um die einfachen Grundideen klar heraustreten zu lassen.

Eine ausführliche, mit vollständigen Beweisen versehene Darlegung wird an anderer Stelle erscheinen.

---

1) Lehrstuhl für Mathematik IV, Universität Mannheim,
Seminargebäude A 5, B 123
D-6800 Mannheim 1

<u>1. Polynomiale Splines mit einfachen Knoten.</u> Es seien m und n natürliche Zahlen mit $m \geq 2$. Wir betrachten ein reelles kompaktes Intervall $[a, b]$. Gegeben seien $n+1$ reelle Zahlen x_ν, die Knoten, mit

$$a = x_0 < x_1 < \cdots < x_{n-1} < x_n = b.$$

Die Menge dieser Knoten heiße K_n.

Eine reelle Funktion $s \in C[a, b]$ heißt eine Splinefunktion bzw. ein Spline der Ordnung m bezüglich der Knotenmenge K_n, wenn die Restriktion von s auf das Teilintervall $[x_{\nu-1}, x_\nu], \nu = 1, 2, \cdots, n,$ dem Polynomraum Π_{m-1} angehört, und wenn für $m > 2$ die Funktion s stetige Ableitungen bis zur Ordnung $m-2$ auf dem Intervall $[a, b]$ besitzt. Der endlich-dimensionale Vektorraum dieser Splines wird mit $S_m(K_n)$ bezeichnet. Es gilt

$$dim S_m(K_n) = m + n - 1.$$

Führt man für $t \in \mathbb{R}$ und $k \in \mathbb{N}$ die "+"-Funktionen (truncated power functions) durch

$$(t)_+^k := \begin{cases} 0 & \text{für } t < 0, \\ t^k & \text{für } t \geq 0 \end{cases}$$

ein, so kann man für den Raum $S_m(K_n)$ leicht eine Basis angeben. In Form einer direkten Summe gilt

$$S_m(K_n) = \Pi_{m-1} \oplus span\left\{(x - x_\nu)_+^{m-1}; \nu = 1, 2, \cdots, n-1\right\}. \tag{1}$$

Für viele theoretische und praktische Zwecke ist dagegen eine aus sog. B-Splines gebildete Basis günstiger. Wir erweitern der Einfachheit halber hierzu die Knotenmenge und betrachten unendlich viele Knoten $x_\nu, \nu \in \mathbb{Z}$, mit

$$x_\nu < x_{\nu+1},$$

die sich im Endlichen nicht häufen. Die zugehörige Knotenmenge heiße K.

Zu gegebenem Index $\nu \in \mathbb{Z}$ gibt es genau eine auf $\mathbb{R}$ definierte reelle Funktion, den sog. B-Spline $B_{m,\nu}(x)$, die den folgenden Bedingungen genügt: $B_{m,\nu}$ ist eine Splinefunktion der Ordnung m auf dem Intervall $[x_\nu, x_{\nu+m}]$; sie verschwindet identisch für $x \leq x_\nu$ und für $x \geq x_{\nu+m}$ und besitzt auf $\mathbb{R}$ stetige Ableitungen bis zur Ordnung $m-2$; ferner ist

$$\int_{\mathbb{R}} B_{m,\nu}(x) dx = 1.$$

Es stellt sich heraus, daß

$$B_{m,\nu}(x) > 0$$

ist für $x \in (x_\nu, x_{\nu+m})$.

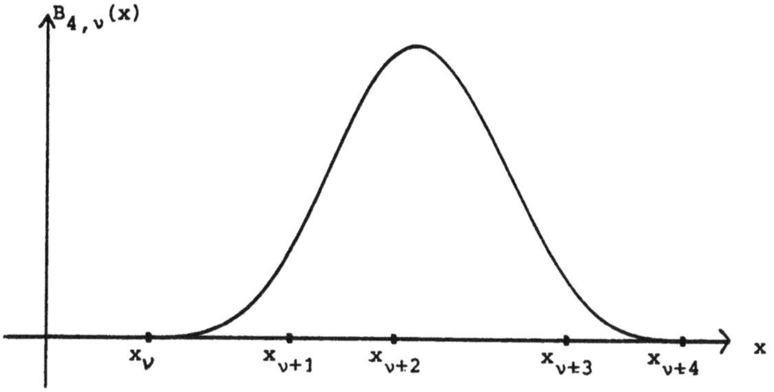

Fig.1: Graph von $B_{4,\nu}(x)$.

Ferner gilt für den ursprünglichen Spline-Raum

$$S_m(K_n) = span\,\{B_{m,\nu};\nu = -m+1, -m+2, \cdots, n-1\}\,. \tag{2}$$

Die folgende Rekursionsformel erlaubt es, mit B-Splines in elementarer Weise zu verfahren (vgl. [3],[4],[11],[13]): Für alle $\nu \in \mathbb{Z}$ und $m \geq 3$ gilt für $x \in \mathbb{R}$ die Rekursion

$$B_{m,\nu}(x) = \frac{m}{m-1}\left\{\left(\frac{x - x_\nu}{x_{\nu+m} - x_\nu}\right)B_{m-1,\nu}(x) + \left(\frac{x_{\nu+m} - x}{x_{\nu+m} - x_\nu}\right)B_{m-1,\nu+1}(x)\right\}. \tag{3}$$

Für $m = 2$ ist

$$B_{2,\nu}(x) = \frac{2}{(x_{\nu+2} - x_{\nu+1})(x_{\nu+2} - x_\nu)(x_{\nu+1} - x_\nu)}\{(x_{\nu+2} - x_{\nu+1})(x - x_\nu)_+ -$$

$$-(x_{\nu+2} - x_\nu)(x - x_{\nu+1})_+ + (x_{\nu+1} - x_\nu)(x - x_{\nu+2})_+\}$$

$$= \begin{cases} 0 & \text{für } x \leq x_\nu, \\ \frac{2(x-x_\nu)}{(x_{\nu+2}-x_\nu)(x_{\nu+1}-x_\nu)} & \text{für } x \in [x_\nu, x_{\nu+1}], \\ \frac{2(x_{\nu+2}-x)}{(x_{\nu+2}-x_\nu)(x_{\nu+2}-x_{\nu+1})} & \text{für } x \in [x_{\nu+1}, x_{\nu+2}], \\ 0 & \text{für } x \geq x_{\nu+2}. \end{cases}$$

Für spätere Zwecke geben wir noch eine Darstellung von $B_{m,\nu}$ mit Hilfe eines komplexen Kurver integrals an (vgl. [12]): Es sei $\mathbb{R}$ die reelle Achse der komplexen z-Ebene. Für $x \in \mathbb{R}$ sei C_x ein einfach geschlossene, rektifizierbare Kurve der z-Ebene, so daß alle Knoten $x_\mu \in K$ mit

$$x < x_\mu \leq x_{\nu+m}$$

im Innern von C_x liegen. Dann gilt

$$B_{m,\nu}(x) = \frac{1}{2\pi i}\int\limits_{C_x} \frac{m(z-x)^{m-1}}{(z-x_\nu)(z-x_{\nu+1})\cdots(z-x_{\nu+m})}dz\,, \tag{4}$$

wobei im positiven Sinne integriert wird.

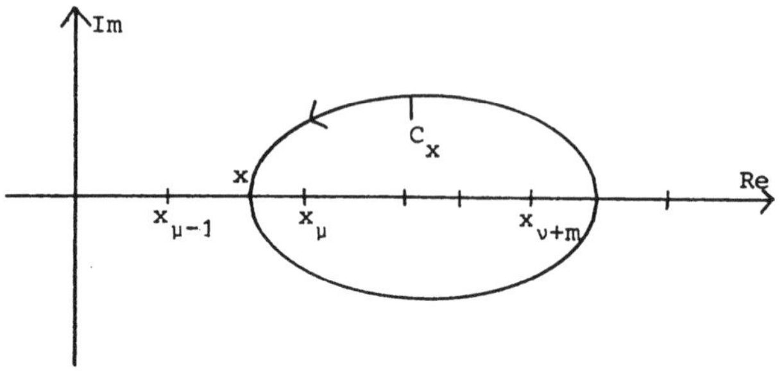

Fig. 2: Zum Integrationsweg C_x in (4).

Die Integraldarstellung (4) erlaubt problemlos den Übergang von einfachen Knoten zu solchen mit Vielfachheiten > 1, denn in (4) kann man die Voraussetzung fallen lassen, daß die Knoten paarweise verschieden sind. Lediglich die Bedingung

$$x_\nu < x_{\nu+m}$$

muß erfüllt bleiben, um für die modifizierte Funktion $B_{m,\nu}$ einen Träger positiver Länge zu garantieren.

Ein reeller Teilraum V von $C[a,b]$ von endlicher Dimension,

$$dim\ V = r \in \mathbb{N} \ ,$$

erfüllt die Haar'sche Bedingung, wenn jede von der Nullfunktion verschiedene Funktion $v \in V$ höchstens $r-1$ Nullstellen im Intervall $[a,b]$ besitzt. In diesem Fall ist das Problem der numerischen Konstruktion der hier eindeutigen besten Tschebyscheff-Approximation relativ leicht lösbar.
Offenbar erfüllt der Raum $S_m(K_n)$ diese Bedingung nicht, sofern $n > 1$ ist, wie das Beispiel

$$s(x) = (x - x_1)^{m-1} - (x - x_1)_+^{m-1} = \begin{cases} (x - x_1)^{m-1} & \text{für } x \in [a, x_1], \\ 0 & \text{für } x \in [x_1, b] \end{cases}$$

zeigt.
Ein reller Teilraum V von $C[a,b]$ von endlicher Dimension,

$$dim\ V = r \in \mathbb{N} \ ,$$

erfüllt die schwache Haar'sche Bedingung, wenn jede Funktion $v \in V$ höchstens $r-1$ Vorzeichenwechsel im Intervall $[a,b]$ besitzt, d.h. wenn es zu gegebenem v keine $r+1$ Punkte ξ_μ mit

$$a \le \xi_1 < \xi_2 < \ldots < \xi_{r+1} \le b$$

gibt, so daß

$$v(\xi_\mu)v(\xi_{\mu+1}) < 0$$

ist für $\mu = 1, 2, \ldots, r$.

Der Raum $S_m(K_n)$ erfüllt in der Tat die schwache Haar'sche Bedingung ([5], vgl. auch [11] und [13]). Diese Eigenschaft ermöglicht via Störung die Konstruktion von Räumen, die zum Raum $S_m(K_n)$ isomorph sind und die Haar'sche Bedingung erfüllen.

2. Gauß-Transformation von Spline-Räumen. Für $x \in [a,b]$ kann jede Splinefunktion $s \in S_m(K_n)$ wegen (2) eindeutig in der Form

$$s(x) = \sum_{\nu=-m+1}^{n-1} \alpha_\nu B_{m,\nu}(x) \tag{5}$$

dargestellt werden. Auf Grund der Definition der B– Splines kann für x in (5) jede reelle Zahl eingesetzt werden. Die dadurch entstehende Fortsetzung von s auf $\mathbb{R}$ heiße s_1. Nun ordnen wir jeder Splinefunktion $s \in S_m(K_n)$ mit einem Parameter $t > 0$ die Gauß-Transformierte

$$u(x,t) := \frac{1}{\sqrt{4\pi t}} \int_{-\infty}^{+\infty} s_1(\tau) \exp(-\frac{(\tau - x)^2}{4t}) d\tau \tag{6}$$

zu. Für jedes $x \in \mathbb{R}$ gilt

$$\lim_{t \to 0_+} u(x,t) = s_1(x) . \tag{7}$$

Ferner erfüllt $u(x,t)$ für $x \in \mathbb{R}, t > 0$, die Wärmeleitungsgleichung

$$u_t = u_{xx} . \tag{8}$$

Der mit Hilfe von (6) entstehende Bildraum heiße $G_m(K_n,t)$. Es gilt

$$dim\ G_m(K_n,t) = m + n - 1 .$$

Ferner erfüllt $G_m(K_n,t)$ die Haar'sche Bedingung. Die Elemente von $G_m(K_n,t)$ sind Restriktionen ganzer Funktionen der Wachstumsordnung 2 auf $\mathbb{R}$.

Eine Basis von $G_m(K_n,t)$ besteht aus den Bildern der B– Splines unter der Gauß-Transformation, d.h. aus den Funktionen

$$F_{m,\nu}(x,t) := \frac{1}{\sqrt{4\pi t}} \int_{-\infty}^{+\infty} B_{m,\nu}(\tau) \exp(-\frac{(\tau - x)^2}{4t}) d\tau \tag{9}$$

für $\nu = -m+1, -m+2, \ldots, n-1$. Definiert man noch

$$F_{1,\nu}(x,t) := \frac{1}{x_{\nu+1} - x_\nu} \frac{1}{\sqrt{4\pi t}} \int_{x_\nu}^{x_{\nu+1}} \exp(-\frac{(\tau - x)^2}{4t}) d\tau \tag{10}$$

und

$$F_{0,\nu}(x,t) := \frac{1}{\sqrt{4\pi t}} \exp(-\frac{(x-x_\nu)^2}{4t}) , \tag{11}$$

so kann gezeigt werden, daß für alle $m \geq 2$ die Rekursion

$$\left.\begin{aligned} F_{m,\nu}(x,t) = {} & \frac{m}{(m-1)(x_{\nu+m}-x_\nu)} \{(x-x_\nu)F_{m-1,\nu}(x,t) + (x_{\nu+m}-x)F_{m-1,\nu+1}(x,t)\} \\ & + \frac{2tm}{(x_{\nu+m}-x_\nu)} \left\{ \frac{F_{m-2,\nu}(x,t)}{x_{\nu+m-1}-x_\nu} - \left(\frac{1}{x_{\nu+m-1}-x_\nu} + \frac{1}{x_{\nu+m}-x_{\nu+1}} \right) \cdot \right. \\ & \left. \cdot F_{m-2,\nu+1}(x,t) + \frac{F_{m-2,\nu+2}(x,t)}{x_{\nu+m}-x_{\nu+1}} \right\} \end{aligned}\right\} \tag{12}$$

gilt. Somit können die Basisfunktionen $F_{m,\nu}$ elementar und stabil ausgewertet werden, sofern nur für die Funktionen $F_{1,\nu}$ gute Approximationen zur Verfügung stehen. Dies ist aber der Fall, weil $F_{1,\nu}$ mit dem Fehlerintegral

$$\operatorname{erf}(x) = \phi(x) = \frac{2}{\sqrt{\pi}} \int_0^x e^{-\tau^2} d\tau = 1 - \operatorname{erfc}(x) \quad , \quad x \geq 0 , \tag{13}$$

in Verbindung steht. Es ist

$$F_{1,\nu}(x,t) = \frac{1}{2(x_{\nu+1}-x_\nu)} \begin{cases} \left(\phi\left(\frac{x_{\nu+1}-x}{\sqrt{4t}}\right) - \phi\left(\frac{x_\nu-x}{\sqrt{4t}}\right)\right) & \text{für } x \leq x_\nu, \\ \left(\phi\left(\frac{x_{\nu+1}-x}{\sqrt{4t}}\right) + \phi\left(\frac{x-x_\nu}{\sqrt{4t}}\right)\right) & \text{für } x_\nu \leq x \leq x_{\nu+1}, \\ \left(-\phi\left(\frac{x-x_{\nu+1}}{\sqrt{4t}}\right) + \phi\left(\frac{x-x_\nu}{\sqrt{4t}}\right)\right) & \text{für } x \geq x_{\nu+1}. \end{cases}$$

Ferner sind Approximationen für die Funktion $\operatorname{erfc}(x)$ mit hoher Genauigkeit verfügbar. Bei mittlerer Präzision gilt z.B.

$$\operatorname{erfc}(x) = e^{-x^2} \frac{P(x)}{Q(x)} + \delta(x)$$

mit

$$P(x) = 1.0000001 + 0.6698148x + 0.1625179x^2 ,$$

$$Q(x) = 1 + 1.7382060x + 1.1236830x^2 + 0.2831084x^3$$

und

$$\max_{0 \leq x < \infty} |\delta(x)| \leq 1.3 \cdot 10^{-7} .$$

<u>3. Gleichmäßige Approximation mit Splinefunktionen.</u> Zu einer gegebenen reellen Funktion $f \in C[a,b]$ wird eine Splinefunktion $s \in S_m(K_n)$ gesucht, für die die Tschebyscheff-Norm

$$\|f - s\| := \max_{x \in [a,b]} |f(x) - s(x)|$$

möglichst klein ist. Eine beste Approximation ist i.a. nicht eindeutig bestimmt. Ferner ist das simultane Austauschverfahren von Remez nicht ohne erhebliche Modifikation anwendbar (vgl. [13]). Daher betrachten wir für festes $t > 0$ die gleiche Aufgabe im transformierten Raum $G_m(K_n,t)$.

Hier kann der effiziente Remez-Algorithmus zur Konstruktion der eindeutig bestimmten besten Approximation herangezogen werden. Diese beste Approximation $g(x,t)$ kann in der Form

$$g(x,t) = \sum_{\nu=-m+1}^{n-1} c_\nu(t) F_{m,\nu}(x,t)$$

dargestellt werden. Um zur Spline-Approximation zu gelangen, verwenden wir eine heuristische Strategie. Sie beruht darauf, daß man ein asymptotisches Verhalten

$$c_\nu(t) = c_\nu^* + t d_\nu + o(t)$$

mit Konstanten c_ν^*, d_ν für $t \to 0$ als richtig voraussetzt. Berechnen wir nun die beste Approximation bezüglich des Raumes $G_m(K_n, t_0)$ und bezüglich des Raumes $G_m(K_n, \frac{t_0}{2})$, so setzt man im Sinn der sog. Richardson-Extrapolation

$$\hat{c}_\nu := 2c_\nu\left(\frac{t_0}{2}\right) - c_\nu(t_0)$$

für $\nu = -m+1, -m+2, \ldots, n-1$. Betrachtet wird die Splinefunktion

$$s(x) := \sum_{\nu=-m+1}^{n-1} \hat{c}_\nu B_{m,\nu}(x)$$

als Approximation an die Funktion f. Die Güte der Approximation kann durch Berechnung oberer und unterer Schranken für die Minimalabweichung a posteriori beurteilt werden. Gegebenenfalls muß durch weitere Halbierung des Parameters eine neue Approximation bzgl. $G_m(K_n, \frac{t_0}{4})$ berechnet werden, wobei sich in bekannter Weise weitere Verfeinerungen der Extrapolation anbieten usw. Möglicherweise können zur Stützung bzw. zur Verbesserung einer solchen Strategie Eigenschaften des durch die Wärmeleitungsgleichung gelieferten Halbgruppen-Operators eingesetzt werden.

4. Lösung einer Anfangswertaufgabe der Wärmeleitungsgleichung. In einfacher Weise kann man mit Hilfe unseres Formelsystems (10), (11), (12) die folgende Aufgabe in leicht auswertbarer Form lösen. Es sei eine reelle Funktion $f \in C[0,1]$ mit $f(0) = f(1) = 0$ gegeben. Gesucht ist eine zweimal stetig differenzierbare Funktion $u(x,t)$ auf dem Halbstreifen

$$0 < x < 1 \quad , \quad 0 < t < \infty ,$$

die dort der Wärmeleitungsgleichung

$$u_t = u_{xx}$$

genügt, stetig auf dem Bereich

$$0 \leq x \leq 1 \quad , \quad 0 \leq t < \infty$$

ist und dort die Anfangswerte

$$u(x,0) = f(x) \quad \text{für} \quad 0 \leq x \leq 1$$

und die Randwerte

$$u(0,t) = u(1,t) = 0 \quad \text{für} \quad 0 \leq t < \infty$$

annimmt. Bekanntlich ist die einzige – wachstumsbeschränkte – Lösung dieser Aufgabe für $t > 0$ durch das Integral

$$u(x,t) = \frac{1}{\sqrt{4\pi t}} \int_{-\infty}^{+\infty} \tilde{f}(\tau) \exp(-\frac{(\tau - x)^2}{4t}) d\tau \tag{14}$$

gegeben. Hier ist die Funktion $\tilde{f}$ durch die ungerade periodische Fortsetzung von f der Periode 2 auf $\mathbb{R}$ definiert, d.h.

$$\tilde{f}(x) := \begin{cases} f(x) & \text{für } x \in [0,1], \\ -f(-x) & \text{für } x \in [-1,0], \\ \tilde{f}(x+2) & \text{für alle } x \in \mathbb{R}. \end{cases}$$

Um das Integral mit Hilfe des rekursiven Verfahrens auswerten zu können, müssen wir $\tilde{f}$ durch eine Summe von $B-$ Splines annähern. Hierzu führen wir im Spline-Raum $S_m(K_n)$, der für $a = 0, b = 1$ betrachtet wird, einige Manipulationen durch. Zunächst besteht K_n aus den Knoten $x_0, x_1, \ldots, x_n$ mit

$$0 = x_0 < x_1 < x_2 < \ldots < x_n = 1 \; .$$

Die Knoten werden am Nullpunkt gespiegelt, d.h. wir definieren

$$x_\nu := -x_{-\nu} \quad \text{für} \quad \nu = -1, -2, \ldots, -n \; .$$

Dann werden sie mit der Periode 2 periodisch auf $\mathbb{R}$ fortgesetzt, d.h. für Indizes $\mu, \nu \in \mathbb{Z}$ mit

$$\mu = \nu + r \cdot 2n \quad , \quad r \in \mathbb{Z}$$

sei

$$x_\mu = x_\nu + 2r \; .$$

Die Menge dieser so entstandenen Knoten heiße $\tilde{K}$. Der Spline-Raum der Ordnung m zu dieser Knotenmenge heiße $S_m(\tilde{K})$. Der Untervektorraum $P_m(\tilde{K})$ bestehe aus den periodischen Splines $s \in S_m(\tilde{K})$ der Periode 2. Es gilt (vgl. [11]):

$$dim\; P_m(\tilde{K}) = 2n \; .$$

Eine große Zahl von Interpolationsproblemen mit periodischen Splines aus $P_m(\tilde{K})$ kann auf einfache Weise gelöst werden. Dies gilt z.B. für gerade Ordnung m und die Interpolation an den Knoten. Die Lösung des Interpolationsproblems kann wieder als Summe von – periodisch fortgesetzten – $B-$ Splines dargestellt werden.

Interpoliert bzw. approximiert man nun die gegebene Funktion $\tilde{f}$ unter Erhaltung der Bedingungen

$$\tilde{f}(0) = \tilde{f}(1) = 0$$

durch Splines aus $P_m(\tilde{K})$, so gewinnt man die Lösung des Anfangswertproblems als Linearkombination der periodisch bzgl. x fortgesetzten Basisfunktionen $F_{m,\nu}(x,t)$ für $t > 0$. Der Rekursionsprozess (10), (11), (12) stellt sicher, daß, abgesehen vom Approximations- bzw. Interpolationsfehler keine weiteren Verfahrensfehler mehr bei der Auswertung der Lösung auftreten.

5. Ein Vergleichssatz. Wegen (1) kann jede Splinefunktion $s \in S_m(K_n)$ eindeutig in der Form

$$s(x) = p(x) + \sum_{\nu=1}^{n-1} \beta_\nu (x - x_\nu)_+^{m-1} \tag{15}$$

mit $p \in \Pi_{m-1}$ dargestellt werden. In (15) ist es auf der rechten Seite offenbar erlaubt, für x jede reelle Zahl einzusetzen. Die sich hierdurch ergebende Fortsetzung von s auf $\mathbb{R}$ werde mit s_2 bezeichnet. Sie ist i.a. von der früher via $B-$ Splines erfolgten Fortsetzung s_1 von s verschieden.

Wir unterwerfen s wieder der Gauß-Transformation: Jedem $s \in S_m(K_n)$ wird für $t > 0$ die Funktion

$$v(x,t) := \frac{1}{\sqrt{4\pi t}} \int_{-\infty}^{+\infty} s_2(\tau) \exp\left(-\frac{(\tau - x)^2}{4t}\right) d\tau \tag{16}$$

zugeordnet. Es gilt für jedes $x \in \mathbb{R}$ die Beziehung

$$\lim_{t \to 0+} v(x,t) = s_2(x) \,. \tag{17}$$

Der Bildraum von $S_m(K_n)$ unter der Transformation (16) heiße $W_m(K_n, t)$. Er erfüllt auf dem Intervall $[a, b]$ die Haar'sche Bedingung. Es handelt sich sogar dort um einen sog. Polya-Raum (vgl. [11]). Zur Gewinnung einer Basis von $W_m(K_n, t)$ definieren wir die spezielle Funktion – eine Transformation der " + " – Funktion –

$$w(x,t) := \frac{1}{\sqrt{4\pi t}} \int_0^{\infty} \tau^{m-1} \exp\left(-\frac{(\tau - x)^2}{4t}\right) d\tau \tag{18}$$

In Analogie zu (1) erhalten wir die direkte Summendarstellung

$$W_m(K_n, t) = \Pi_{m-1} \oplus span\,\{w(x - x_\nu, t); \nu = 1, 2, \cdots, n-1\}\,. \tag{19}$$

Hier geht der Raum Π_{m-1} bei der Transformation (16) in sich über.
Für jede Funktion $y \in W_m(K_n, t)$ gilt: Es gibt eindeutig bestimmte reelle Zahlen γ_ν, so daß

$$D^m y := \frac{d^m y}{dx^m} = \sum_{\nu=1}^{n-1} \gamma_\nu \exp\left(-\frac{(x - x_\nu)^2}{4t}\right)$$

ist. Wir verwenden weiter die Abkürzung D für den Operator der Differentiation. Dann ergibt sich: Der Raum $W_m(K_n, t)$ ist der Lösungsraum der linearen homogenen Differentialgleichung der Ordnung $m + n - 1$:

$$\left(D - \frac{x_{n-1}}{2t}\right)\left(D - \frac{x_{n-2}}{2t}\right) \cdots \left(D - \frac{x_1}{2t}\right) \exp\left(\frac{x^2}{4t}\right) D^m y = 0 \,. \tag{20}$$

Durch leichte Umformung ergibt sich der Operator L, definiert durch:

$$Ly := (2tD + x - x_{n-1})\,(2tD + x - x_{n-2}) \cdots (2tD + x - x_1)\,D^m y \tag{21}$$

und die zu (20) äquivalente Differentialgleichung

$$Ly = 0.$$

Die Kette von Räumen

$$\Pi_0 \subset \Pi_1 \subset \cdots \subset \Pi_{m-1} \subset W_m(K_2, t) \subset W_m(K_3, t) \subset \cdots \subset W_m(K_n, t)$$

von jeweils um 1 zunehmender Dimension bildet ein sog. reguläres Haar'sches System (vgl. [8], [9], [10]) bzw. eine Kette von Polya-Räumen (vgl. [11]). Somit gilt der Vergleichssatz:

Es sei $t > 0$. Die beiden Funktionen f und g seien auf $[a,b]$ mindenstens $(m+n-1)$-mal stetig differenzierbar. Es gelte

$$|(Lf)(x)| \le (Lg)(x) \tag{22}$$

für alle $x \in [a,b]$. Dann gilt für die Minimalabweichungen bei der Approximation bzgl. des Raumes $W_m(K_n, t)$ die Ungleichung

$$\rho_{W_m}(f) \le \rho_{W_m}(g). \tag{23}$$

Für $t \to 0$, d.h. beim Übergang zum Spline-Raum $S_m(K_n)$, reduziert sich der Operator L zu

$$L_0 := (x - x_{n-1})(x - x_{n-2}) \cdots (x - x_1) D^m.$$

Offenbar treten Schwierigkeiten bei der Erfüllung der Ungleichung (22) auf. In diesem Zusammenhang wird auf die Resultate in [1],[2] und in [14] hingewiesen.

Literatur

[1] BRAESS,D.: On the Degree of Approximation by Spline Functions with Free Knots. Aequationes math.12,80-81,1975.

[2] BRAESS,D.: Nonlinear Approximation Theory. Springer-Verlag,Berlin,Heidelberg,New York,1986.

[3] COX,M.G.: The Numerical Evaluation of B-Splines. J.Inst.Math.Applic.10,134-149,1972.

[4] deBOOR,C.: On Calculating with B-Splines. J.Approx.Th.6,50-62,1972.

[5] JONES,R.C.and L.A.KARLOVITZ: Equioscillation under Nonuniqueness in the Approximation of Continuous Functions.J.Approx.Th.3,138-145,1970

[6] KAISER,U.: Das Schoenberg'sche Approximationsproblem. Dissertation, Mannheim,1987.

[7] KARLIN,S.and W.J.STUDDEN: Tchebycheff-Systems with Applications in Analysis und Statistics.Interscience Publ.,New York,London,1966.

[8] MEINARDUS,G.: Über ein Monotonieprinzip bei linearen Approximationen. ZAMM 46,227-238,1966.

[9] MEINARDUS,G.: Approximation of Functions: Theory and Numerical Methods. Springer-Verlag,Berlin,Heidelberg,New York,1967.

[10] MEINARDUS,G.: Zur Abschätzung der Minimalabweichung bei linearer Approximation. ZAMM 50,509-514,1970.

[11] MEINARDUS,G.: Approximation Theory. In Vorbereitung. Springer-Verlag,Berlin,Heidelberg,New York.

[12] MEINARDUS,G.: Bemerkungen zur Theorie der B-Splines.Erschienen in: Spline-Funktionen(Hrsg.Böhmer,Meinardus,Schempp),Mannheim,1974.

[13] NÜRNBERGER,G.: Approximation by Spline-Functions. In Vorbereitung.Springer-Verlag,Berlin,Heidelberg,New York.

[14] STRAUSS,H.: Monotonicity of Quadrature Formulae of Gauss Type and Comparison Theorems for Monosplines.Numer.Math.44,337-347,1984.

COMPUTATION OF INNER PRODUCTS OF SOME MULTIVARIATE SPLINES

Edward Neuman

Department of Mathematics
Southern Illinois University
Carbondale, Illinois 62901-4408
U.S.A.

1. Introduction. In some problems that arise in projection methods, including the finite element method and least squares, we often need to compute the inner product of multivariate splines (or of some functions related to them). Among different classes of multivariate splines the so-called box splines play an important role in several different contexts. These splines have been introduced by de Boor and De Vore [1], and systematically studied in [2,3,7,8,9] (see also references therein). Box splines can be viewed as generalizations of univariate B-splines with equally spaced knots (cf. [2]).

The outline of this paper is as follows. In Section 2 we introduce basic notation and also present some auxiliary results. In Section 3 we show that the inner product of derivatives (partial or directional) of multivariate box splines equals the value of an appropriate box spline at some point. Also we show that the inner product of two functions under discussion can be expressed in terms of the backward difference of a cone spline. In Section 4 we specialize the general results of the previous section to the case of univariate box splines as well as to the case of multivariate box splines on an s+1-direction mesh in $\mathbb{R}^s$ ($s \geq 2$). The latter class of box splines is of special interest in the theory of finite element method. We close this section with some applications of our previous results to the evaluation of multiple integrals involving powers of the sinc function.

Let us note that some methods for the exact evaluation of inner products of multivariate simplex splines are discussed in [6] and [11].

2. Basic notation and lemmas. Let us now introduce some notation and definitions which will be used throughout the sequel. By x,y,..., we denote elements of Euclidean space $\mathbb{R}^s$ ($s \geq 1$), i.e. $x = (x_1,\ldots,x_s)^T$. Superscripts are used to number vectors. The Euclidean inner product of $x,y \in \mathbb{R}^s$ is denoted by $x \cdot y$. For a given set $A \subset \mathbb{R}^s$, $\chi_A(x)$, $[A]$, $\langle A \rangle$, and $|A|$ mean the characteristic function of A, the convex hull, the linear span, and the cardinality of A, respectively. We use standard multi-index notation, i.e. for $\alpha \in \mathbb{Z}_+^s$, $|\alpha| = \alpha_1 + \ldots + \alpha_s$, $\alpha! = \alpha_1! \cdots \alpha_s!$. Furthermore, we set

$$D^\alpha = \left[\frac{\partial}{\partial x_1}\right]^{\alpha_1} \cdots \left[\frac{\partial}{\partial x_s}\right]^{\alpha_s}.$$

Let $Z = \{z^1,\ldots,z^n\} \subset \mathbb{R}^s\backslash\{0\}$ ($n \geq s \geq 1$). To this end we will assume that $\langle Z \rangle = \mathbb{R}^s$. The box spline with the knot set Z, denoted by $B(\cdot|Z)$, is defined as in [1] by requiring that

$$\int_{\mathbb{R}^s} f(x)B(x|Z)dx = \int_{[0,1]^n} f(t_1z^1 + \ldots + t_nz^n)dt_1 \cdots dt_n$$

for any $f \in C(\mathbb{R}^s)$.

A useful number associated with $B(\cdot|Z)$ is

$$d \equiv d(Z) := \operatorname{Max}\{r: \langle Z\backslash U \rangle = \mathbb{R}^s \text{ for all } U \subseteq Z \text{ with } |U| = r\}.$$

It is known that

$$B(\cdot|Z) \in C^{d(Z)-1}(\mathbb{R}^s).$$

For our further purposes we define $\delta := z^1 + \ldots + z^n$ and

$$\Phi(y|Z) := \prod_{j=1}^{n} S(y \cdot z^j/2), \qquad (y \in \mathbb{R}^s)$$

where

$$S(t) = \frac{\sin t}{t}$$

is the sinc function. Further, let

$$\hat{f}(y) = \int_{\mathbb{R}^s} e^{-iy\cdot x} f(x)dx$$

denote the Fourier transform of $f \in L_2(\mathbb{R}^s)$. By the inversion formula

$$f(x) = (2\pi)^{-s} \int_{\mathbb{R}^s} e^{iy\cdot x} \hat{f}(y)dy.$$

Our first auxiliary result is contained in the following.

Lemma 2.1. Let $c \in \mathbb{R}^s$. Then

$$B(x-c|Z) = (2\pi)^{-s} \int_{\mathbb{R}^s} e^{iy\cdot(x-c-\delta/2)} \Phi(y|Z)dy. \tag{2.1}$$

Proof. It is well known that

$$\hat{B}(y-c|Z) = e^{-iy\cdot c} \prod_{j=1}^{n} \frac{1-e^{-iy\cdot z^j}}{iy\cdot z^j} \tag{2.2}$$

(cf. [2]). Hence

$$\hat{B}(y-c|Z) = e^{-iy\cdot c} \prod_{j=1}^{n} e^{-iy\cdot z^j/2} S(y\cdot z^j/2)$$

or

$$\hat{B}(y-c|Z) = e^{-iy\cdot(c+\delta/2)} \Phi(y|Z). \tag{2.3}$$

Applying the inversion formula to (2.3), we obtain the assertion.

□

Before the next formula let us introduce more notation. For $f \in C(\mathbb{R}^s)$ and for $v \in \mathbb{R}^s$, the backward-difference operator ∇_v is given by

$$(\nabla_v f)(\cdot) := f(\cdot)-f(\cdot-v).$$

Also we set

$$\nabla_Z := \prod_{v\in Z} \nabla_v .$$

Throughout this paper the symbol $T(\cdot|Z)$ stands for the cone spline with knot set Z. Assume $0 \notin [Z]$. Then the cone spline is defined by requiring that

$$\int_{\mathbb{R}^s} f(x)T(x|Z)dx = \int_{\mathbb{R}^n_+} f(t_1z^1 + \ldots + t_nz^n)dt_1 \cdots dt_n$$

for any $f \in C_0(\mathbb{R}^s)$ (see [4]). It is known that

$$(2.4) \qquad \hat{T}(y|Z) = \prod_{j=1}^{n} \frac{1}{iy\cdot z^j}.$$

In the next lemma a simple relationship between box splines and cone splines is given. This result is essentially due to W. Dahmen and C. A. Micchelli [7]. The proof is included for the completeness of presentation.

Lemma 2.2 ([7]). Under the above assumptions

$$(2.5) \qquad B(x-c|Z) = \nabla_Z T(x-c|Z) \qquad (c \in \mathbb{R}^s).$$

Proof. It follows from (2.2) and (2.4) that

$$\hat{B}(y-c|Z) = e^{-iy\cdot c}\, \hat{T}(y|Z) \prod_{j=1}^{n} (1-e^{-iy\cdot z^j}).$$

Since

$$\prod_{j=1}^{n} (1-e^{-iy\cdot z^j}) = (\nabla_Z e^{iy\cdot x})(0),$$

$$\hat{B}(y-c|Z) = e^{-iy\cdot c}\, \hat{T}(y|Z)[(\nabla_Z e^{iy\cdot x})(0)]$$
$$= e^{-iy\cdot c}[\nabla_Z \hat{T}(\cdot|Z)](0).$$

Hence

$$\hat{B}(y-c|Z) = \nabla_Z \hat{T}(y-c|Z).$$

Applying the inversion formula, we obtain immediately the desired result. □

3. Main results. In this section we give closed formulas for the inner products of the partial derivatives and the directional derivatives of box splines. Let

$$X = (x^1,\dots,x^l) \quad (l \geq s \geq 1), \quad Y = (y^1,\dots,y^m) \quad (m \geq s \geq 1)$$

be subsets of $\mathbb{R}^s\backslash\{0\}$. We will assume that $\langle X\rangle = \mathbb{R}^s$ and $\langle Y\rangle = \mathbb{R}^s$. Further, let

$$(3.1) \qquad Z := X \cup Y = (x^1,\dots,x^l,y^1,\dots,y^m) \equiv (z^1,\dots,z^n)$$

where $n = l+m$. We let $\delta_1 = x^1 + \dots + x^l$, $\delta_2 = y^1 + \dots + y^m$,

$\delta = \delta_1+\delta_2$. Also we need the following functions:

$$\Phi_1(y|X) \equiv \prod_{j=1}^{l} S(y\cdot x^j/2),$$

$$\Phi_2(y|Y) \equiv \prod_{j=1}^{m} S(y\cdot y^j/2), \text{ and}$$

$$\Phi(y|Z) \equiv \Phi_1(y|X)\Phi_2(y|Y) \qquad (y \in \mathbb{R}^s).$$

The inner product utilized here is the one given by

$$\langle f,g\rangle := \int_{\mathbb{R}^s} f(x)\bar{g}(x)dx,$$

($f,g \in L_2(\mathbb{R}^s)$) where as usual $\bar{g}$ denotes the complex conjugate of g. For the sake of notation we will write $I^{\alpha,\beta}(a,X;b,Y)$ instead of $\langle D^\alpha B(\cdot-a|X),\ D^\beta B(\cdot-b|Y)\rangle$ ($\alpha,\beta \in \mathbb{Z}_+^s$, $a,b \in \mathbb{R}^s$).

We are now in a position to state and prove the following.

Theorem 3.1. Let $\alpha,\beta \in \mathbb{Z}_+^s$ be such that $|\alpha| \le d(X)-1$ and $|\beta| \le d(Y)-1$. Then

$$(3.1) \qquad \begin{aligned} I^{\alpha,\beta}(a,X;b,Y) &= (-1)^{|\beta|}[D^{\alpha+\beta}B(\cdot|Z)](b-a+\delta_2) \\ &= (-1)^{|\alpha|}[D^{\alpha+\beta}B(\cdot|Z)](a-b+\delta_1) \end{aligned}$$

($a,b \in \mathbb{R}^s$). Moreover, if $0 \notin [Z]$, then

$$(3.1) \qquad \begin{aligned} I^{\alpha,\beta}(a,X;b,Y) &= (-1)^{|\beta|}[\nabla_Z D^{\alpha+\beta}T(\cdot|Z)](b-a+\delta_2) \\ &= (-1)^{|\alpha|}[\nabla_Z D^{\alpha+\beta}T(\cdot|Z)](a-b+\delta_1). \end{aligned}$$

Proof. We take advantage of the fact that the box spline is a rapidly decreasing function. Thus

$$\widehat{D^\alpha B}\ (y-a|X) = (iy)^\alpha \hat{B}(y-a)|X).$$

This in conjuction with (2.3) yields

$$(3.3) \qquad \widehat{D^\alpha B}\ (y-a|X) = (iy)^\alpha e^{-iy\cdot(a+\delta_1/2)}\Phi_1(y|X).$$

The same arguments applied to $D^\beta B(\cdot-b|Y)$ give

$$(3.4) \qquad \widehat{D^\beta B}\ (y-b|y) = (iy)^\beta e^{-iy\cdot(b+\delta_2/2)}\Phi_2(y|Y).$$

Next making use of Parseval's identity, we arrive at

$$I^{\alpha,\beta}(a,X;b,Y) = (2\pi)^{-s}\langle\widehat{D^\alpha B}\ (\cdot-a|X),\widehat{D^\beta B}\ (\cdot-b|Y)\rangle.$$

Hence (3.3) and (3.4) yield

(3.5) $I^{\alpha,\beta}(a,X;b,Y) =$

$$= (-1)^{|\beta|} \frac{i^{|\alpha+\beta|}}{(2\pi)^s} \int_{\mathbb{R}^s} y^{\alpha+\beta} e^{iy(b-a-\delta_1/2+\delta_2/2)} \Phi_1(y|X)\Phi_2(y|Y)\,dy.$$

On the other hand, (2.1) gives

$$D^{\alpha+\beta}B(x-c|Z) = \frac{i^{|\alpha+\beta|}}{(2\pi)^s} \int_{\mathbb{R}^s} y^{\alpha+\beta} e^{iy\cdot(x-c-\delta/2)} \Phi(y|Z)\,dy.$$

Inserting $x = b-a+c+\delta_2$ and $\delta = \delta_1+\delta_2$ into the last formula, we obtain in conjunction with (3.5) the first part of (3.1). The second formula of (3.1) follows from the previous one and from the fact that $I^{\alpha,\beta}(a,X;b,Y) = I^{\beta,\alpha}(b,Y;a,X)$. For the proof of (3.2) we employ (2.5) to obtain

$$D^{\alpha+\beta}B(x-c|Z) = \nabla_Z D^{\alpha+\beta}T(x-c|Z).$$

To complete the proof we set the above $x = b-a+c+\delta_2$ (or $x = a-b+c+\delta_1$) and then apply (3.1). □

In the next theorem we give closed formulas for the inner products of the directional derivatives of box splines.

For $u \in \mathbb{R}^s$, the directional derivative of $f \in C^1(\mathbb{R}^s)$ in the direction u is given by

$$(D_u f)(x) = \sum_{j=1}^{s} u_j \frac{\partial f(x)}{\partial x_j}.$$

The result reads as follows.

Theorem 3.2. Let $u,v \in \mathbb{R}^s$. If

$$D_u B(\cdot-a|X), D_v B(\cdot-b|Y) \in L_2(\mathbb{R}^s),$$

then

$$\text{(3.6)} \quad \begin{aligned} \langle D_u B(\cdot-a|X), D_v B(\cdot-b|Y)\rangle &= -[D_u D_v B(\cdot|Z)](b-a+\delta_2) \\ &= -[D_u D_v B(\cdot|Z)](a-b+\delta_1) \end{aligned}$$

Moreover, if $0 \notin [Z]$, then

$$\text{(3.7)} \quad \begin{aligned} \langle D_u B(\cdot-a|X), D_v B(\cdot-b|Y)\rangle &= -[D_u D_v(\nabla_Z T(\cdot|Z))](b-a+\delta_2) \\ &= -[D_u D_v(\nabla_Z T(\cdot|Z))](a-b+\delta_1). \end{aligned}$$

Proof. Let $f \in C^1(\mathbb{R}^s)$. Then

$$\int_{\mathbb{R}^s} B(x-a|X)(D_u f)(x)dx = \sum_{j=1}^{s} u_j \int_{\mathbb{R}^s} B(x-a|X)\frac{\partial}{\partial x_j} f(x)dx.$$

Applying integration by parts to the last integral we obtain

$$(3.8) \qquad \int_{\mathbb{R}^s} B(x-a|X)(D_u f)(x)dx = \int_{\mathbb{R}^s} \text{div } F(x)dx - \int_{\mathbb{R}^s} f(x) D_u B(x-a|X)dx,$$

where

$$F(x) = \{u_1 B(x-a|X)f(x),\ldots,u_s B(x-a|X)f(x)\}$$

is the vector field. Since supp $B(\cdot-a|X)$ is a bounded subset in $\mathbb{R}^s$, there is a bounded, closed $\Omega \subset \mathbb{R}^s$ such that supp $B(\cdot-a|X) \subseteq \Omega$. Taking into account that $B(x-a|X) \equiv 0$ if $x \in \delta\Omega$ (the boundary of Ω) and applying the divergence theorem, we obtain

$$\int_{\mathbb{R}^s} \text{div } F(x)dx = 0.$$

Hence from this result and (3.8) we have

$$\int_{\mathbb{R}^s} f(x) D_u B(x-a|X)dx = -\int_{\mathbb{R}^s} B(x-a|X)(D_u f)(x)dx.$$

Inserting $f(x) = e^{-iy\cdot x}$ into the last formula, we arrive at

$$\widehat{D_u B}\,(y-a|X) = (iy\cdot u)\hat{B}(y-a|X).$$

By the same arguments

$$\widehat{D_v B}\,(y-b|Y) = (iy\cdot v)\hat{B}(y-b|Y).$$

The remainder of the proof follows the lines introduced in the proof of Theorem 3.1. □

4. Applications. In this section we specialize our previous results to the case of univariate box splines (§4.1) and multivariate box splines on an s+1-direction mesh in $\mathbb{R}^s$ (§4.2). Also some formulas for the definite integrals involving the sinc function are included (§4.3).

4.1. Let $s = 1$. We assume that the knot sets X, Y, and Z are the same as those in Section 3. Throughout this subsection we will write $I^{p,q}(a,X;b,Y)$ for the inner product of $B^{(p)}(\cdot-a|X)$ and $B^{(q)}(\cdot-b|Y)$ $(p,q \in \mathbb{Z}_+)$.

We have the following.

Proposition 4.1. Let $a,b \in \mathbb{R}$ and let $p,q \in \mathbb{Z}_+$ where $p \leq l-1$, $q < m-1$. If $z^j > 0$ $(z^j < 0)$ for all $j = 1,2,\ldots,n$, then

$$I^{p,q}(a,X;b,Y) = (-1)^q c_1 (\nabla_Z x_+^{n-\nu-1})(b-a+\delta_2) \tag{4.1}$$
$$= (-1)^p c_1 (\nabla_Z x_+^{n-\nu-1})(a-b+\delta_1)$$

where $\nu = p+q$, $x_+^t = (\mathrm{Max}\,\{0,x\})^t$ $(t \geq 0)$,

$$c_1 = \left[(n-\nu-1)!\left[\prod_{j=1}^{l} x^j\right]\left[\prod_{j=1}^{m} y^j\right]\right]^{-1} \tag{4.2}$$

$(\delta_1 = x^1 + \ldots + x^l,\ \delta_2 = y^1 + \ldots + y^m)$.

Proof. Since all the z's are positive (negative), $0 \notin [Z]$. In such a case the univariate cone spline $T(\cdot|Z)$ can be expressed in the following way.

$$T(x|Z) = \left[(n-1)!\left[\prod_{j=1}^{l} x^j\right]\left[\prod_{j=1}^{m} y^j\right]\right]^{-1} x_+^{n-1}$$

(see [11]). From this and (3.2), the assertion follows. □

The above result can be applied for the evaluation of inner products of the derivatives of univariate B-splines with equally spaced knots. To this aim let us choose $x^1 = \ldots = x^l = h_1$, $y^1 = \ldots = y^m = h_2$ with $h_1 h_2 > 0$. It is known ([2]) that with these assumptions

$$B(\cdot-a|X) = M(\cdot|a,a+h_1,\ldots,a+lh_1) \equiv M_1(\cdot)$$

and

$$B(\cdot-b|Y) = M(\cdot|b,b+h_2,\ldots,b+mh_2) \equiv M_2(\cdot).$$

Here M_1 and M_2 are the univariate B-splines of order l and m, respectively with the knot sets as shown above. Using this and Proposition 4.1 we obtain immediately the following.

Corollary 4.2. Let $p,q \in \mathbb{Z}_+$ ($p \le l-1$, $q \le m-1$). If $h_1 h_2 > 0$, then

$$\langle M_1^{(p)}, M_2^{(q)} \rangle = (-1)^q c_2 (\nabla_Z x_+^{n-\nu-1})(b-a+mh_2)$$
$$= (-1)^p c_2 (\nabla_Z x_+^{n-\nu-1})(a-b+lh_1)$$

where $c_2 = [(n-\nu-1)!\, h_1^l h_2^m]^{-1}$ ($n = l+m$, $\nu = p+q$),

$$Z = \{\underbrace{h_1,\dots,h_1}_{l},\ \underbrace{h_2,\dots,h_2}_{m}\}.$$

In particular, when $h_1 = h_2 \equiv h$, then

$$\langle M_1^{(p)}, M_2^{(q)} \rangle = (-1)^q c_2 \sum_{j=0}^{n} (-1)^j \binom{n}{j} [b-a+(m-j)h]_+^{n-\nu-1}$$
$$= (-1)^p c_2 \sum_{j=0}^{n} (-1)^j \binom{n}{j} [a-b+(l-j)h]_+^{n-\nu-1}. \qquad \square$$

It follows from the last formula that for any $0 \le p \le l-1$

$$\|M^{(p)}\|_2^2 = \frac{(-1)^p}{(2l-2p-1)!\, h^{2p+1}} \sum_{j=1}^{l} (-1)^{l-j} \binom{2l}{l+j} j^{2l-2p-1}$$

where now $M(\cdot)$ is the univariate B-spline (of order l) with the mesh size equal to h.

4.2. Assume now $s > 1$. By e_j ($1 \le j \le s$) we will denote the jth coordinate vector in $\mathbb{R}^s$. Also let $e_0 = (1,\dots,1) \in \mathbb{R}^s$.

An $s+1$-direction mesh in $\mathbb{R}^s$ is a set of vectors

$$X = \{\underbrace{e_0,\dots,e_0}_{l_0},\dots,\underbrace{e_s,\dots,e_s}_{l_s}\}, \tag{4.3}$$

where $l_j \ge 1$, $j = 0,1,\dots,s$. It is clear that $\langle X \rangle = \mathbb{R}^s$. Thus there exists a unique box spline $B(\cdot|X)$.

For our purposes we define also

$$(4.4) \qquad Y = \{\underbrace{e_0,\dots,e_0}_{m_0},\dots,\underbrace{e_s,\dots,e_s}_{m_s}\}$$

with $m_j \geq 1$, all j. Further, as in Section 3 we set

$$(4.5) \qquad Z := X \cup Y = \{\underbrace{e_0,\dots,e_0}_{\nu_0},\dots,\underbrace{e_s,\dots,e_s}_{\nu_s}\}$$

where $\nu_j = l_j+m_j$ $(0 \leq j \leq s)$. For the remainder of this section let $n := \nu_0 + \dots + \nu_s$ and $\nu := (\nu_0,\dots,\nu_s)$. Since $0 \notin [Z]$, there exists a unique cone spline $T(\cdot|Z)$.

By $I(a,X;b,Y)$ we will denote the inner product of $B(\cdot-a|X)$ and $B(\cdot-b|Y)$.

We are ready to state and prove the following.

Proposition 4.2. Let $a,b \in \mathbb{R}^s$. Then

$$(4.6) \qquad I(a,X;b,Y) = v_Z T(b-a+\delta_2|Z) = v_Z T(a-b+\delta_1|Z),$$

where X, Y, and Z are given by (4.3)-(4.5), respectively, $\delta_2 = m_0e_0 + \dots + m_se_s$, $\delta_1 = l_0e_0 + \dots + l_sm_s$. If Z is given by (4.5), then

$$(4.7) \qquad T(x|Z) = \begin{cases} \gamma_{\nu,s}\displaystyle\int_{c_s}^{d_s}(\omega-\rho)^{\nu_0-1}\prod_{j=1}^{s}[\rho+(s-1)x_j-\omega]^{\nu_j-1}\,d\rho & \text{if } x > 0, \\ 0 & \text{otherwise.} \end{cases}$$

Here $x = (x_1,\dots,x_s) \in \mathbb{R}^s$, $\omega = x_1 + \dots + x_s$,

$$(4.8) \qquad \gamma_{\nu,s} = [(\nu_0-1)! \dots (\nu_s-1)!(s-1)^{n-s-1}]^{-1},$$

$$(4.9) \qquad c_s = \mathrm{Max}\,\{x_j : 1 \leq j \leq s\}, \text{ and}$$

$$(4.10) \qquad d_s = \mathrm{Min}\,\{x_j + x_k : 1 \leq j < k \leq s\}.$$

Proof. The first assertion is a special case of (3.2). Since supp $T(\cdot|Z) = \{x \in \mathbb{R}^s: x_j \geq 0, 1 \leq j \leq s\}$ (the cone spanned by the vectors $e_0,\ldots,e_s$), we need to establish the first part of (4.7), only. Assume $x_j > 0$, for all j. For our purposes we employ the result

$$T(x|Z) = \frac{1}{(n-1)!}\int_0^\infty \tau^{-n+s-1}M(\tau x|Z)d\tau \tag{4.11}$$

(cf. [11],[5]). Here $M(\cdot|Z)$ denotes the simplex spline with the knot set Z, which in the case under discussion takes the form

$$M(x|Z) = (n-1)!c_\nu\chi_A(x)\prod_{j=0}^{s}[\lambda_j(x)]^{\nu_j-1} \tag{4.12}$$

(see [7]). Here

$$c_\nu = \left[\prod_{j=0}^{s}(\nu_j-1)!\right]^{-1}, \tag{4.13}$$

A denotes an s simplex spanned by $e_0,\ldots,e_s$, and $\lambda_0(x),\ldots,\lambda_s(x)$ are the barycentric coordinates of the point $x \in \mathbb{R}^s$, i.e.,

$$e_0\lambda_0(x) + \ldots + e_s\lambda_s(x) = x, \text{ with}$$
$$\lambda_0(x) + \ldots + \lambda_s(x) = 1.$$

Solving this linear system one obtains

$$\lambda_0(x) = (\omega-1)/(s-1), \text{ and}$$
$$\lambda_j(x) = [1+(s-1)x_j-\omega]/(s-1), \quad (1 \leq j \leq s).$$

Combining this with (4.12) and (4.13) yields

$$M(\tau x|Z) = (n-1)!\gamma_{\nu,s}\chi_A(\tau x)(\tau\omega-1)^{\nu_0-1}\prod_{j=1}^{s}[1+(s-1)x_j-\tau\omega]^{\nu_j-1},$$

where $\gamma_{\nu,s}$ is given by (4.8). Inserting this into (4.11), one obtains

$$T(x|Z) = \gamma_{\nu,s}\int_0^\infty \chi_A(\rho^{-1}x)(\omega-\rho)^{\nu_0-1}\prod_{j=1}^{s}[\rho+(s-1)x_j-\omega]^{\nu_j-1}d\rho. \tag{4.14}$$

Since $x > 0$, $\rho^{-1}x \in A$ iff

$$0 \le \rho^{-1}x_j \le 1 \quad (1 \le j \le s)$$

and

$$\rho^{-1}(x_{\mu_1}+\dots+ x_{\mu_p}) \ge 1 \quad (1 \le \mu_1 \le\dots\le \mu_p \le s,\ 2 \le p \le s).$$

Solving with respect to ρ, we obtain

$$\rho \ge \mathrm{Max}\ \{x_j\colon 1 \le j \le s\} =: c_s$$

and

$$\rho \le \mathrm{Min}\ \{x_{\mu_1}+\dots+x_{\mu_p}\colon 1 \le \mu_1 \le\dots\le \mu_p,\ \ 2 \le p \le s\}.$$

Taking into account that $x_j > 0$, for all j, one obtains

$$\rho \le \mathrm{Min}\ \{x_j+x_k\colon 1 \le j < k \le s\} =: d_s.$$

Thus

$$\chi_A(\rho^{-1}x) = \begin{cases} 1 & c_s \le \rho \le ds, \\ 0 & \text{otherwise.} \end{cases}$$

Combining this with (4.14) gives the first line of (4.7). This completes the proof. □

The integral that appears in the right-hand side of (4.7) can be computed exactly with the aid of the Newton-Cotes quadrature formulas (the integrand is an algebraic polynomial of degree n-s-1 or less). Analyzing the formula (4.6) we see that the number of the different cone splines that appear on the right-hand side of (4.6) is not bigger than 2^n.

4.3. We close this paper with some formulas for the integrals involving powers of the sinc function $S(t) = \frac{\sin t}{t}$. Let

$$Z = \{z^1,\dots,z^n\} \subset \mathbb{R}^s\setminus\{0\}$$

where $n \ge s \ge 1$. We assume that $\langle Z\rangle = \mathbb{R}^s$. Setting $x = c+\delta/2$ ($c \in \mathbb{R}^s$) into (2.1) and substituting $y = 2t$ ($t \in \mathbb{R}^s$) we obtain

$$(4.15) \qquad \int_{\mathbb{R}^s} \prod_{j=1}^{n} S(t\cdot z^j)dt = \pi^s B(\delta/2|Z)$$

where $\delta = z^1 +\dots+ z^n$.

Let the knot set Z be given by (4.5). Then in view of Lemma 2.2 and (4.15)

$$\int_{\mathbb{R}^s}\left[\frac{\sin(t_1+\ldots+t_s)}{t_1+\ldots+t_s}\right]^{\nu_0}\left[\frac{\sin t_1}{t_1}\right]^{\nu_1}\cdots\left[\frac{\sin t_s}{t_s}\right]^{\nu_s}dt_1\cdots dt_s$$

$$= \pi^s[\nabla_Z T(\cdot|Z)](\delta/2)$$

where now $\delta = (\nu_0+\nu_1,\ldots,\nu_0+\nu_s)^T$. The cone splines $T(\cdot|Z)$ can be evaluated with the aid of (4.7).

Also from (4.15) for $s = 1$ and $Z = \underbrace{\{1,\ldots,1\}}_{n}$

$$\int_{\mathbb{R}}\left[\frac{\sin t}{t}\right]^n dt = \pi\, B\left[\frac{n}{2}\Big|Z\right] = \pi\, M\left[\frac{n}{2}\Big|0,1,\ldots,n\right]$$

where $M(\cdot|0,1,\ldots,n)$ denotes the univariate B-spline with knots at $0,1,\ldots,n$. Hence also

$$\int_{\mathbb{R}}\left[\frac{\sin t}{t}\right]^n dt = \frac{\pi}{(n-1)!}\sum_{j=w}^{n}(-1)^{n-j}\binom{n}{j}\left[j-\frac{n}{2}\right]^{n-1}$$

where

$$w = \begin{cases}(n+2)/2 & n\text{-even},\\ (n+1)/2 & n\text{-odd}.\end{cases}$$

The last result seems to be little known.

REFERENCES

[1] C. de Boor and R. De Vore, Approximation by smooth multivariate splines, Trans. Amer. Math. Soc. 276(1983), pp. 775-788.

[2] C. de Boor and K. Höllig, B-splines from parallelepipeds, J. d.'Anal. Math. 42(1982/83), pp. 99-115.

[3] ________, Bivariate box splines and pp functions on a three direction mesh, J. Comput. Appl. Math 9(1983), pp. 13-28.

[4] W. Dahmen, Multivariate B-splines-Recurrence relations and linear combinations of truncated powers, in Multivariate Approximation Theory, W. Schempp, K. Zeller, eds., Birkhäuser Verlag, Basel, 1979, pp. 64-82.

[5] W. Dahmen and C. A. Micchelli. On limits of multivariate B-splines, J. d.'Anal. Math. 39(1981), pp. 256-278.

[6] ________, Computation of inner products of multivariate B-splines, Numer. Funct. Anal. Optim. 3(1981), pp. 367-375.

[7] ________, Recent progress in multivariate splines, in Approximation Theory IV, C. K. Chui, L. L. Schumaker, and J. D. Ward, eds., Academic Press, New York, 1983, pp. 27-121.

[8] ________, Translates of multivariate splines, Lin. Alg. Appl. 52/53(1983), pp. 217-234.

[9] ________, On the optimal approximation rates for criss-cross finite element spaces, J. Comput. Appl. Math. 10(1984), pp. 255-273.

[10] T. A. Grandine, The evaluation of inner products of multivariate simplex splines, SIAM J. Numer. Anal. 24(1987), pp. 882-886.

[11] C. A. Micchelli, On numerically efficient method for computing B-splines, in Multivariate Approximation Theory, W. Schempp and K. Zeller, eds., Birkhäuser Verlag, Basel, 1979, pp. 211-248.

ON ESTIMATES FOR ONE-DIMENSIONAL SPLINE APPROXIMATION

P. Oswald [1)]

In this note we deal with function space techniques for one-dimensional spline approximation theory. In Section 1 we introduce the usual Besov-Sobolev spaces on an interval. In Section 2 a scale of approximation spaces based on spline spaces with respect to a certain increasing sequence of quasi-uniform partitions is defined. Here as well as in Section 3 dealing with the direct and inverse theorems of spline approximation theory we follow some ideas developed in our previous papers [15] [16]. In Section 4 we formulate some results on the degree of nonlinear spline approximation and briefly discuss their relationships to the h-version of the finite element method.
To a great extent, the approach presented can be extended to higher dimensions (cf. [15], [16], [8] for first results in this direction).

1. Classical function spaces

Let $I = [\alpha, \beta]$ be some finite interval. By $C^m(I)$, $m = 0,1,\ldots$, we denote the Banach spaces of m times continuously differentiable functions on I equipped with the maximum norm, and by $L_p(I)$ the quasi-Banach spaces of (equivalence classes of) p-integrable functions on I equipped with the quasi-norm

$$\|f\|_{L_p(I)} = \begin{cases} \left(\int_I |f(x)|^p \, dx\right)^{1/p}, & 0 < p < \infty \\ \operatorname*{ess\,sup}_{x \in I} |f(x)|, & p = \infty. \end{cases}$$

The Sobolev-Lebesgue spaces $W_p^m(I)$, $1 \leq p \leq \infty$, $m = 1,2,\ldots$, are defined by

$$W_p^m(I) = \left\{ f \in C^{m-1}(I) : f^{(m-1)} \text{ is absolutely continuous, } f^{(m)} \in L_p(I) \right\}$$

$$\|f\|_{W_p^m(I)} = \|f\|_{L_p(I)} + \|f^{(m)}\|_{L_p(I)}$$

Furthermore, set $W_p^0(I) = L_p(I)$, $1 \leq p \leq \infty$.

1) P. Oswald, Sektion Mathematik, FSU Jena, Universitätshochhaus, Jena, 6900, DDR

For introducing the Besov spaces $B^{s,m}_{p,q}(I)$ we require the so-called L_p-moduli of smoothness

$$\omega^m(t,f)_p = \sup_{0\le h\le t} \|\Delta^m_h f\|_{L_p(I)} \quad , \ 0<p\le\infty, \ m = 1,2,\ldots$$

where

$$\Delta^m_h f(x) = \begin{cases} \sum_{k=0}^{m} \binom{m}{k}\cdot(-1)^{m-k}\cdot f(x+kh) & , \text{ if } x, x+mh \in I \\ 0 \ , & \text{elsewhere.} \end{cases}$$

Then

$$B^{s,m}_{p,q}(I) = \left\{ f\in L_p(I) \ : \ |f|_{B^{s,m}_{p,q}(I)} = \left\| \{2^{is}\cdot\omega^m(2^{-i},f)_p\}_{i\ge 0} \right\|_{l_q} < \infty \right\}$$

where l_q denotes the usual sequence space, $0<q\le\infty$, $s>0$.

Equipped with the quasi-norm

$$\|f\|_{B^{s,m}_{p,q}(I)} = \|f\|_{L_p(I)} + |f|_{B^{s,m}_{p,q}(I)}$$

the Besov spaces form a scale of non-trivial quasi-Banach spaces where $m = 1,2,\ldots$, $0<p,q\le\infty$, $0<s\le m+\delta$ with $\delta = \max(0,1/p-1)$ (in the case $s = m+\delta$ the Besov space will be non-trivial only for $q = \infty$)
There is a lot of alternative definitions which are, however, all equivalent to the above given explicit inner description via the L_p-moduli of smoothness (cf. e.g. [22], [23]), at least for the range $\delta < s < m$.
Finally, let $H^m_p(I) = W^m_p(I)$, $m = 0,1,\ldots$, and define for non-integer $s>0$ and $1<p<\infty$

$W^s_p(I) = B^{s,m}_{p,p}(I)$ (Sobolev-Slobodeckii spaces, $s<m$)

$H^s_p(I) = [W^m_p, W^{m'}_p]_\Theta$ (Lebesgue spaces, $m<s<m'$, $\Theta = \frac{s-m}{m'-m}$

two further scales of function spaces frequently used in the theory of differential and integral equations. Here $[\cdot,\cdot]_\Theta$ denotes the complex interpolation method between Banach spaces. For these definitions which are independent (up to equivalent norms) of the choice of m, m' and more detailed information on function spaces and interpolation theory we refer to [24], [5].
Although most of theoretical numerical analysis is based on Sobolev-Lebesgue space methos, especially in the cases $p = 2$ where $H^s(I) \equiv H^s_2(I) = W^s_2(I) = B^{s,m}_{2,2}(I)$ for $0<s<m$, and $p = \infty$, there is an increasing interest in other quasinorms both for applications to nonlinear problems and for improved solution procedures for particular linear problems. Since we are mainly dealing with approximation estimates and some of their applications to error estimation

for spline methods, the Besov spaces are most appropriate, see also [25], [26], [3]. Moreover, if approximation results are available in Besov norms then, as a rule, they automatically yield estimates for Sobolev-Lebesgue norms by using the elementary imbeddings

$B^{s,m}_{p,1}(I) \subsetneq W^s_p(I)$, $H^s_p(I) \subsetneq B^{s,m}_{p,\infty}(I)$, $0 < s < m$.

We conclude this Section by considering an example which will be used for illustration. Let $I = [0,1]$ and $f(x) = x^\gamma$, $\gamma \neq$ integer. Such functions model (in the one-dimensional setting) the singularities of solutions of boundary value problems in plane polygonal domains (cf. [11]). A straightforward calculation shows that for $t \to 0$

$$\omega^m(t,f)_p \asymp \begin{cases} t^m & , \gamma > m - 1/p \\ t^m \cdot |\log t|^{1/p}, & \gamma = m - 1/p \\ t^{\gamma+1/p} & , -1/p < \gamma < m - 1/p \ . \end{cases}$$

Thus, x^γ is of maximal smoothness $s = m$ (i.e. $x^\gamma \in B^{s,m}_{p,\infty}(I)$) iff $p < (m-\gamma)^{-1}$.

2. Spline and approximation spaces

Let $\pi = \{I_j = (x_{j-1}, x_j),\ j = 1,\dots,M\}$ be some partition of I into M intervals where $\alpha = x_0 < x_1 < \dots < x_{M-1} < x_M = \beta$, and set $h = h(\pi) = \max |I_j|$, $\underline{h} = \underline{h}(\pi) = \min |I_j|$ where $|I|$ denotes the length of an interval I.

Let $m = 1,2,\dots$, $r = 1,\dots,m$. By $S^{m,r}(\pi)$ we denote the finite-dimensional subspace of $C^{m-r-1}(I)$ ($L_\infty(I)$ if $r = m$) of polynomial splines of degree $m - 1$ and of defect r with respect to π, i.e. $g \in S^{m,r}(\pi)$ iff the restrictions of g to the intervals I_j are polynomials of degree $\leq m - 1$, and $g \in C^{m-r-1}(I)$ if $r < m$ resp. $g \in L_\infty(I)$ if $r = m$. The most important cases of this definition are $r = 1$ (smooth splines), $r = m-1$ resp. $m-2$ (one-dimensional C^0-resp. C^1-finite element functions), and $r = m$ (piecewise polynomial functions).

As in [16] we introduce now a certain scale of approximation spaces ($s > 0$, $0 < p, q \leq \infty$, m, r as above). Consider any increasing sequence of dyadic, quasi-uniform partitions $\pi_i = \{I_{i,j},\ j = 1,\dots,M_i\}$, i.e. satisfying

(1) $\pi_0 \prec \pi_1 \prec \pi_2 \prec \dots$ where $\pi \prec \pi'$ means that any interval of π' is contained in some interval belonging to π.

(2) $0 < C_1 \leq 2^i \cdot h_i \leq C_2 < \infty$, $h_i = h(\pi_i)$, $i \geq 0$

(3) $h_i/\underline{h}_i \leq C_3$, $i \geq 0$

(in the following the positive constants C_k, $k = 1,2,3$, will preserve this meaning while by C we denote some generic constant).

A function $f \in L_p(I)$ belongs by definition to $A^{s,m,r}_{p,q} \equiv$

$A^s_{p,q}(\{S^{m,r}(\pi_i)\}_{i\geq 0})$ if

(4) there exists a sequence $g_i \in S^{m,r}_i \equiv S^{m,r}(\pi_i)$, $i \geq 0$

such that $f = \sum_{i\geq 0} g_i$ in the sense of $L_p(I)$ and

$$\| \{2^{is} \cdot \|g_i\|_{L_p(I)}\}_{i\geq 0} \|_{l_q} < \infty .$$

The representation (4) which is not unique will be called admissible. Then, by

(5) $$\|f\|_{A^{s,m,r}_{p,q}} = \inf \| \{2^{is} \cdot \|g_i\|_{L_p(I)}\}_{i\geq 0} \|_{l_q}$$

where the infimum is taken with respect to all admissible representations (4) we define the quasi-norm in $A^{s,m,r}_{p,q}$.

We first quote some properties of these spaces which are consequences of the considerations in [15], [16] (cf. also [8]).

<u>Theorem 1 .</u> Let $m = 1,2,\ldots$, $r = 1,\ldots,m$, $0 < p, q \leq \infty$, and $s > 0$ be given.

(a) $A^{s,m,r}_{p,q}$ is a quasi-Banach space (Banach space iff $1 \leq p, q \leq \infty$).

(b) Let $E^{m,r}_i(f)_p = \inf\{\|f - g\|_{L_p(I)} : g \in S^{m,r}_i\}$ be the best approximation of $f \in L_p(I)$ by the subspace $S^{m,r}_i$, and consider any sequence $g^*_i \in S^{m,r}_i$ of quasi-best approximating elements, i.e.

$$\|f - g^*_i\|_{L_p(I)} \leq C_4 \cdot E^{m,r}_i(f)_p , \quad i \geq 0.$$

Then

$$\|f\|^*_{A^{s,m,r}_{p,q}} = \|f\|_{L_p(I)} + \| \{2^{is} \cdot E^{m,r}_i(f)_p\}_{i\geq 0} \|_{l_q}$$

and

$$\| \{2^{is} \cdot \|g^{**}_i\|_{L_p(I)}\}_{i\geq 0} \|_{l_q} , \quad g^{**}_i = \begin{cases} g^*_0, & i = 0 \\ g^*_i - g^*_{i-1}, & i > 0 \end{cases}$$

form equivalent quasi-norms on $A^{s,m,r}_{p,q}$ (in the latter case with constants depending on C_4 but not on the concrete choice of the sequence g^*_i).

(c) $B^{s,m}_{p,q}(I) \hookrightarrow A^{s,m,r}_{p,q}$ (continuous imbedding)

(d) $A^{s,m,r}_{p,q} \subsetneq B^{s,m}_{p,q}(I)$ if $0<s<m-r+1/p$

(e) $A^{s,m,r}_{p,q} = B^{s,m}_{p,q}(I)$ if $0<s<m-r+1/p$ with equivalent quasi-norms.

(f) The definition makes sense for $s=0$, $q \leq \min(1,p)$, too. Then

$$A^{0,m,r}_{p,q} = \begin{cases} L_p(I), & 0<p<\infty, \\ C(I), & p=\infty \text{ and } r \leq m-1,\ m>1 \end{cases}$$

with coinciding quasi-norms.

(The bounds for the imbedding operators and the equivalent quasi-norms are generally depending on the parameters p,q,s,m,r and the constants C_k, $k=1,2,3$, but not on the concrete choice of the underlying sequence $\{\pi_i\}$).

Proof (outline). (b) is implicitly contained in the proof of Theorem 1 in [15] (cf. also [12,ch.5.6], [8]). (a), i.e. the completeness of the spaces follows from (b) as indicated in [12,ch.5.6] for the classical case of trigonometric approximation. The imbedding relations (c) and (d) are consequences of the Jackson and Bernstein type inequalities for spline approximation (see Proposition 1 and Theorem 1 in [15] where the original papers are cited as well). (e) follows from (c), (d). For the proof of (f) it suffices to apply the usual quasi-norm inequalities and the density of $\bigcup S^{m,r}_i$ in $L_p(I)$.

Remark 1. In the context of Fourier analysis, quasi-norms as introduced by (4), (5) are called Nikolskij representations (cf. [12], [24], [22]). What concerns characterizations of function spaces by splines there exists a lot of previous research related to Theorem 1. For instance, characterizations of Besov-Lipschitz spaces via best approximations by splines were considered in [18] ($1 \leq p \leq \infty$) and [14] ($p<1$). The investigation of spline systems as (unconditional) Schauder bases in Besov-Hardy-Sobolev spaces also led to various new equivalent quasi-norms (cf. e.g. [6], [21]). In the author's opinion, the use of the Nikolskij representations and a systematic investigation of the spaces $A^s_{p,q}(\{S_i\})$ (in dependence on the properties of the underlying sequence of subspaces $\{S_i\}$) is of independent interest and might have some important applications in approximation theory and numerical error estimation. For some related results on spline representations for function spaces on R^N, see [20].

3 Approximation and inverse property of $S^{m,r}(\pi)$

The following results (which are relatively easy to prove for the $A^{s,m,r}_{p,q}$ spaces) state the so-called (global) approximation property and the inverse property of $S^{m,r}(\pi)$ which are frequently used in the analysis of spline methods.

Theorem 2 (approximation property). Let $m = 1,2,\dots,\ r = 1,\dots,m$ $0 < p,\ q,\ q' \leq \infty$ and $0 < s < m - r + 1/p,\ s < s' \leq m + \delta$ (with $q' = \infty$ if $s' = m + \delta$) be given. Then, for any $f \in B^{s',m}_{p,q'}(I)$ there exists $g \in S^{m,r}(\pi)$ such that

$$(6)\quad \| f - g \|_{B^{s,m}_{p,q}(I)} \leq C\cdot h^{s'-s}\cdot \| f \|_{B^{s',m}_{p,q'}(I)} .$$

An analogous inequality holds if we replace the B-spaces by the W- or H-spaces for $1 < p < \infty$ (including the case $s = 0$). The constant C in (6) is independent of f and π.

Proof. First, by deleting some knots x_j from the given partition one can construct some smaller partition $\pi^* < \pi$ with the properties $h \leq \underline{h}^* \leq h^* \leq 3\cdot h$ (cf. [14, Lemma 1.1]).

Choosing the integer n according to $2^{-n-1} < h\cdot(\beta-\alpha)^{-1} \leq 2^{-n}$ we can obviously define $\pi_0 < \pi_1 < \dots < \pi_n = \pi^* < \pi_{n+1} < \dots$ in such a way that (2) and (3) are satisfied with some absolute constants C_k, $k = 1,2,3$, independently of π.

Thus, by Theorem 1 (c), (d) for the proof of Theorem 2 it is now sufficient to establish

$$(6)'\quad \| f - \bar{g}_n \|_{A^{s,m,r}_{p,q}} \leq C\cdot 2^{-n(s'-s)}\cdot \| f \|_{A^{s',m,r}_{p,q'}}$$

with some $\bar{g}_n \in S^{m,r}_n \subset S^{m,r}(\pi)$. But (6)' holds even for any $f \in A^{s',m,r}_{p,q'}$ and $0 \leq s < s' < \infty$. For this reason, consider any admissible representation (4) and set

$\bar{g}_n = \sum_{i=0}^{n} g_i \in S^{m,r}_n$. Then, by the definition (5) of the quasinorms

$$\| f - \bar{g}_n \|_{A^{s,m,r}_{p,q}} \leq \| \{2^{is}\cdot \| g_i \|_{L_p(I)}\}_{i>n} \|_{l_q}$$

$$\leq C\cdot 2^{-n(s'-s)}\cdot \| \{2^{is'}\cdot \| g_i \|_{L_p(I)}\}_{i>n} \|_{l_{q'}},$$

where the discrete Hardy inequality has been applied. (6)' follows by taking the infimum with respect to all admissible representations

(The result for the Sobolev-Lebesgue spaces is a consequence of (6) and the imbeddings quoted in Section 1).

Theorem 3 (inverse property). Let $m = 1,2,\dots$, $r = 1,\dots,m$, $0<p,\ q,\ q'\leq\infty$ and $0<s<s'<m-r+1/p$ be given. Then, for any partition π and any $g\in S^{m,r}(\pi)$ we have

$$(7)\quad \|g\|_{B^{s',m}_{p,q'}(I)} \leq C\cdot \underline{h}^{s-s'}\cdot\|g\|_{B^{s,m}_{p,q}(I)} .$$

An analogous inequality holds if we replace the B-spaces by the W- or H-spaces for $1<p<\infty$ (including the case $s = 0$). The constant C in (7) is independent of π and g.

Proof. By including, if necessary, additional knots into π we construct some larger partition $\pi^{**}\succ\pi$ such that π^{**} is quasi-uniform with $\underline{h}/2\leq\underline{h}^{**}\leq h^{**}\leq\underline{h}$ and choose an appropriate sequence $\pi_0\prec\pi_1\prec\dots\prec\pi_n=\pi^{**}\prec\pi_{n+1}\prec\dots$ where n is defined by $2^{-n-1}<\underline{h}\cdot(\beta-\alpha)^{-1}\leq 2^{-n}$ (as above, the constants C_k, $k = 1,2,3$, occuring in (2), (3), will be independent of π).

Thus, in order to establish (7) it remains to prove the corresponding inequality for the $A^{s,m,r}_{p,q}$-spaces in the form

$$(7)'\quad \|g\|^*_{A^{s',m,r}_{p,q'}} \leq C\cdot 2^{n(s'-s)}\cdot\|g\|^*_{A^{s,m,r}_{p,q}},\quad g\in S^{m,r}_n\supset S^{m,r}(\pi)$$

where $0\leq s<s'$ (for $s = 0$ put the $L_p(I)$ quasi-norm in the right-hand side of $(7)'$), see Theorem 1 (b),(c),(d), and (f). But $(7)'$ is obvious:

$$\|g\|^*_{A^{s',m,r}_{p,q'}} \leq \|g\|_{L_p(I)} + \|\{2^{is'}\cdot E^{m,r}_i(g)_p\}_{i<n}\|_{l_{q'}}$$

$$\leq \|g\|_{L_p(I)} + \|\{2^{is}\cdot E^{m,r}_i(g)_p\}_{i<n}\|_{l_\infty}\cdot\|\{2^{i(s'-s)}\}_{i<n}\|_{l_{q'}}$$

$$\leq C\cdot 2^{n(s'-s)}\cdot\|g\|^*_{A^{s,m,r}_{p,\infty}} \leq C\cdot 2^{n(s'-s)}\cdot\|g\|^*_{A^{s,m,r}_{p,q}}$$

(for $s = 0$ observe that $E^{m,r}_i(g)_p\leq\|g\|_{L_p(I)}$)

Remark 2. Theorem 2 and 3 have been established by other methods in [2],[1],[10] for $p = 2$ and the Sobolev-Lebesgue norms. What concerns estimates for particular linear spline approximation schemes (interpolation, Ritz-Galerkin projections), the technique of $A^s_{p,q}$ spaces might be useful, too, because it reduces the considerations to the finite-dimensional subspaces S_i, $i\geq 0$ (see [16] for the case of finite element interpolation in R^2).

4 Nonlinear spline approximation

In this Section we quote some results on the degree of approximation by splines with variable partition from [15] and relate them to the h-version of the finite element method.

Fix a sequence of partitions $\{\pi_i\}$ (satisfying (1),(2),(3)) and the corresponding spline spaces $\{S_j^{m,r}\}$. Let be given some function $f \in A_{p,q}^{s,m,r}$ and some admissible representation (4). Consider, for any $i \geq 0$, the decomposition

$$g_i = \sum_{j=1}^{\bar{M}_i} a_{i,j} \cdot N_{i,j} \qquad (\bar{M}_i = (M_i - 1)\cdot r + m \asymp 2^i \,,\ i \geq 0)$$

with respect to the standard B-spline basis $\{N_{i,j}\}$ in $S_i^{m,r}$ (or some other basis satisfying the L_p-stability condition

$$\|g_i\|_{L_p(I)} \asymp \begin{cases} \left(2^{-i} \cdot \sum_{j=1}^{\bar{M}_i} |a_{i,j}|^p \right)^{1/p}, & 0 < p < \infty \\ \max |a_{i,j}|, & p = \infty . \end{cases}$$

with constants independent of g_i and i, for details cf. [15]).

Now, fix some auxilary constant $C_0 > 0$, some $\varepsilon > 0$ and for any n some integer k such that $n \asymp 2^k$. By

$$(8) \quad \bar{a}_{i,j} = \begin{cases} a_{i,j}, \text{ if } i < k \text{ or if } i \geq k \text{ and } |a_{i,j}| \geq C_0 \cdot 2^{\bar{\alpha} i + \bar{\beta} k} \\ 0, \text{ elsewhere} \end{cases}$$

where $\bar{\alpha} = -s + 1/p + \varepsilon$, $\bar{\beta} = -1/p - \varepsilon$, we define a new set of coefficients. Then, the following properties have been established in [15, Sect. 2]

(9) The number N of non zero coefficients $\bar{a}_{i,j}$ is bounded by

$$N \leq C \cdot 2^k \cdot (1 + C_0^{-p} \cdot \|\{2^{is} \cdot \|g_i\|_{L_p(I)}\}_{i \geq k}\|_{l_q}^p)$$

(10) The function $\bar{g} = \sum_{i \geq 0} \sum_{j=1}^{\bar{M}_i} \bar{a}_{i,j} \cdot N_{i,j}$ defines a spline function belonging to $S^{m,r}(\bar{\pi})$ for some partition $\bar{\pi}$ of I into $\bar{M} \leq C \cdot N$ intervals and satisfying (for sufficiently small ε)

$$\|f - \bar{g}\|_{A_{p',q'}^{s',m,r}} \leq C \cdot 2^{r(s'-s)} \cdot C_0^{1-p/p'} \cdot \|\{2^{is} \cdot \|g_i\|_{L_p(I)}\}\|_{l_q}^{p/p'}$$

$$\|\bar{g}\|_{A^{s,m,r}_{p,q}} \leq \|\{2^{is}\cdot \|g_i\|_{L_p(I)}\}_{i\geq 0}\|_{l_q}$$

where $0<p\leq p'\leq\infty$, $0<q$, $q'\leq\infty$, and $0\leq s'<s-1/p+1/p'$.

Thus, after taking the infimum with respect to all admissible representations (4), the choice $C_0 = \|f\|_{A^{s,m,r}_{p,q}}$ will give

<u>Theorem 4.</u> Let $0<p\leq p'\leq\infty$, $0<q$, $q'\leq\infty$, and $0\leq s'<s-(1/p-1/p')$ be given. Then, for any $f\in A^{s,m,r}_{p,q}$ and for all $n = 1,2,\dots$, there exists a partition $\bar{\pi}_n$ and a spline function $\bar{g}_n\in S^{m,r}(\bar{\pi}_n)$ such that

(11) the dimension of $S^{m,r}(\bar{\pi}_n)$ or, equivalently, the number of intervals in $\bar{\pi}_n$ is bounded by $C\cdot n$,

(12) $\|\bar{g}_n\|_{A^{s,m,r}_{p,q}} \leq C\cdot\|f\|_{A^{s,m,r}_{p,q}}$,

(13) $\|f-\bar{g}_n\|_{A^{s',m,r}_{p',q'}} \leq C\cdot n^{-(s-s')}\cdot\|f\|_{A^{s,m,r}_{p,q}}$.

<u>Remark 3.</u> The constants in (11)-(13) are independent of f and n Moreover, the construction shows that $\bar{\pi}_n$ and $\bar{g}_n$ are "almost" independent of s', p', q' (see (8) and for more details [15]): by a proper choice of $\varepsilon>0$ we get simultaneous approximations for a whole scale of $A^{s',m,r}_{p',q'}$- quasi-norms.

<u>Remark 4.</u> The estimates of Theorem 4 automatically imply analogous results for the scale of Besov-Sobolev spaces (cf. (c),(d) of Theorem 1) In the references given in [15] (**see also the recent papers [17], [7]) one can find some related** information on approximation by splines with free knots.

Let us point out that an application of Theorem 2 together with the usual imbedding relations would only give the estimate,

$$\inf_{g\in S^{m,r}(\bar{\pi}_n)} \|f-g\|_{A^{s',m,r}_{p',q'}} \leq C\cdot n^{-(s-s')+(1/p-1/p')}\cdot\|f\|_{A^{s,m,r}_{p,q}}$$

since $h(\bar{\pi}_n)\gtrsim C\cdot n^{-1}$, see (11). Thus, with respect to the a priori available information $f\in A^{s,m,r}_{p,q}$, the nonlinear procedure of adapting the partition to the individual function f as indicated above yields the better asymptotic rate of approximation (13).

There is an obvious relationship between the above nonlinear approximation problem and such spline methods that involve certain strategies of optimal design of the partition. E.g., Theorem 4 can successfully be used for obtaining appropriate upper estimates of the asymptotic convergence rate of the h-version. For instance, considering the function $f(x) = x^{\gamma}$, $\gamma > 1-1/p'$, $\gamma \neq$ integer, $1 \le p' < \infty$, we have $f(x) \in B^{m,m}_{p,\infty}(I) \hookrightarrow A^{m,m,r}_{p,\infty}$ for $m-\gamma < 1/p$ (see Section 1). On the other hand, we have the imbeddings $W^1_{p'}(I) \hookleftarrow B^{1,m}_{p',1}(I) \supset A^{1,m,r}_{p',1}$ $(m>1,\ m-r+1/p' > 1)$. Thus, Theorem 4 is applicable with $s = m$, $s' = 1$, $q = \infty$, $q' = 1$, $1 \le p' < \infty$ whenever we choose p according to the inequalities $m-\gamma < 1/p < m-1+1/p'$ (which is possible at least for $r = 1,\dots,m-1$, $m>1$). This gives

$$\inf_{\pi:\ M \le n}\ \inf_{g \in S^{m,r}(\pi)} \| x^{\gamma} - g \|_{W^1_{p'}(I)} \le C \cdot n^{-m+1} \cdot \| x^{\gamma} \|_{B^{m,m}_{p,\infty}(I)}$$

(for comparision, for $1-1/p' < \gamma < m-1/p'$ the application of Theorem 2 yields only the weaker $O(n^{-\gamma-1/p'+1})$-estimate). The above inequality is sharp and yields (via the Lax-Milgram and Cea theorems) the asymptotically optimal error estimate in the energy norm achieved by the h-version of the finite element method for second order boundary value problems with solutions including singular parts of power type (we refer to the detailed investigation of this topic for $p' = 2$, $r = m-1$ in [11]).

We conclude this Section with a further possible application of Theorem 4. Let us consider the boundary value problem

(14) $-(a(x)\cdot u'(x))' = f(x)$, $x \in I = [0,1]$, $u(0) = u'(1) = 0$

with right-hand side $f \in L_{p'}(I)$, $1 \le p' < \infty$ and piecewise differentiable coefficient function satisfying $0 < \underline{a} \le a(x) \le \bar{a} < \infty$ on I. Following [13,ch.6] one can prove the existence of a unique generalized solution $u \in V = \{v \in H^1(I):\ v(0) = 0\}$ which coincides with the solution of the weak problem

(14)′ Find $u \in V$ such that for all $v \in V$

$$\int_I a \cdot u' \cdot v'\, dx = \int_I f \cdot v\, dx .$$

and which is piecewise in $W^2_{p'}$ (since u possesses discontinuities in the first derivative it does not belong to $W^2_{p'}(I)$). But by checking the corresponding moduli of smoothness one observes that, at least, $u \in B^{2,2}_{1,\infty}(I)$ and $u' \in B^{1,1}_{1,\infty}(I)$.

Let us now look for the approximate solution of the above problem by a conforming method with linear finite elements. For a given partition π of I, let $V(\pi) = \{g \in S^{2,1}(\pi):\ g(0) = 0\}$ be the finite element space under consideration and denote by $u_{\pi} \in V(\pi)$ the unique solution of the approximate problem

(15) Find $u_\pi \in V(\pi)$ such that for all $v_\pi \in V(\pi)$

$$\int_I a\cdot u'_\pi\cdot v'_\pi \, dx = \int_I f\cdot v_\pi \, dx \ .$$

In the particular case considered we have the simple explicit description

$$u_\pi(x) = \int_0^x w_\pi(t)\, dt \ , \quad x\in I$$

where $w_\pi \in S^{1,1}(\pi)$ is the step function given by

$$w_\pi(x) = \Big(\int_{I_j} a\, dx\Big)^{-1}\cdot \int_{I_j} a\cdot u'\, dx \ , \quad x\in I_j \ , \ j = 1,\dots,M \ .$$

Thus, for arbitrary π we immediately get

$$\|u'_\pi\|_{L_{p'}(I)} \le (\bar a/\underline a)^{1/p'}\cdot \|u'\|_{L_{p'}(I)}$$

as well as

(16) $\|u' - u'_\pi\|_{L_{p'}(I)} \le (1 + (\bar a/\underline a)^{1/p'})\cdot \inf\limits_{\bar w_\pi \in S^{1,1}(\pi)} \|u' - \bar w_\pi\|_{L_{p'}(I)}$

Now, consider some sequence of partitions $\{\pi_i\}$ as introduced in Section 2, and denote $w_i^* = w_{\pi_i} \in S_i^{1,1}$.

Recall that $u' \in B^{1,1}_{1,\infty}(I) \subsetneq A^{1,1,1}_{1,\infty}$ and that according to Theorem 1 (b) we have the admissible representation

$$u' = w_0^* + \sum_{i>0} (w_i^* - w_{i-1}^*) \equiv \sum_{i\ge 0} w_i^{**}$$

and

$$\|\{2^i\cdot \|w_i^{**}\|_{L_1(I)}\}_{i\ge 0}\|_{l_\infty} \le C\cdot \|u'\|_{A^{1,1,1}_{1,\infty}} \le C\cdot \|u'\|_{B^{1,1}_{1,\infty}(I)}$$

(use (16) with $p' = 1$). Thus, we can apply the above construction. Write

$$w_i^{**} = \sum_{j=1}^{M_i} a_{i,j}\cdot 1|_{I_{i,j}} \qquad (a_{i,j} = w_i^{**}|_{I_{i,j}})$$

and set (see (8))

$$\bar a_{i,j} = \begin{cases} a_{i,j}\ , \text{if } i < k \ \text{ or if } i \ge k \ \text{ and } |a_{i,j}| \ge C_0 2^{\bar\alpha i}\cdot n^{\bar\beta} \\[2ex] 0\ , \quad \text{elsewhere} \end{cases}$$

where $\bar\alpha = \varepsilon > 0$, $\bar\beta = -1-\varepsilon$, and $k = [\mathrm{ld}\ n] + 1$. Consider

$$\bar w = \sum_{i\ge 0} \sum_{j=1}^{M_i} \bar a_{i,j}\cdot 1|_{I_{i,j}} = w_k^* + \sum_{i\ge k} \sum_{j=1}^{M_i} \bar a_{i,j}\cdot 1|_{I_{i,j}}$$

and $\bar u(x) = \int_0^x \bar w(t)\, dt \ , \quad x\in I$.

Then, for appropriate choices of the constants C_0 and $\varepsilon > 0$ (e.g., $0 < \varepsilon < 1/(p'-1)$) we obtain by Theorem 4 that $\bar{w} \in S^{1,1}(\bar{\pi})$ and $\bar{u} \in V(\pi) \subset S^{2,1}(\bar{\pi})$ for some partition $\bar{\pi}$ of I into $\leq C \cdot n$ intervals and that

$$\|u - \bar{u}\|_{W^1_{p'}(I)} \leq C \cdot \|u' - w\|_{L_{p'}(I)} \leq C \cdot n^{-1} \cdot \|u'\|_{B^{1,1}_{1,\infty}(I)}$$

Thus, the desired optimal convergence rate can be achieved in this way (for any $p' < \infty$). Observe, however, that the required coefficients $a_{i,j}$ and $\bar{a}_{i,j}$ are only available after the computation of the solutions of a whole sequence of auxiliary approximate problems (15) with $\pi = \pi_i$ $i \geq k$, which is practically impossible or, even when carried out only for a restricted number of partitions π_i with $i \geq k$, too expensive. To avoid this difficulty one could look for strategies of only locally refining the partitions Due to the local character of the approximation scheme it is then possible to obtain the information on the $a_{i,j}$ in some smaller part of I (**with reasonable cossts of** computation). The decision for which part of I further computations should be carried out must be made on the basis of the available a priori and a posteriori information. For our example which models problems with interfaces in the solution (here determined by discontinuities in the coefficient $a(x)$) desirable a priori information would be a lower bound for the distance between neighboured interfaces (but not necessarily the location of the discontinuities) and further smoothness requirements on f. What concerns a posteriori estimators for one-dimensional problems we refer to the theoretical investigations in [4].

Remark 5. In recent time, besides of the h-version, the p- and h-p version of the finite element method attracted much attention. For the approximation theory of the p-version one could introduce an analogous construction. Let P_n be the space of algebraic polynomials of degree $\leq n$ on $I = [0,1]$). Then a function $f \in L_p(I)$ belongs to $A^s_{p,q}(\{P_{2^i}\}_{i \geq 0})$ if

(17) there exists a sequence $p_i \in P_{2^i}$, $i \geq 0$, such that

$f = \sum_{i \geq 0} p_i$ in the sense of $L_p(I)$ and

$\|\{2^{is} \cdot \|p_i\|_{L_p(I)}\}_{i \geq 0}\|_{l_q} < \infty$

with the quasi-norm

$$\|f\|_{A^s_{p,q}(\{p_{2^i}\}_{i\geq 0})} = \inf_{(17)} \|\{2^{is}\cdot \|p_i\|_{L_p(I)}\}_{i\geq 0}\|_{l_q} .$$

From the results stated in [9.ch 7] one obtains characterizations of these spaces by Besov spaces ($p\geq 1$) defined with respect to generalized moduli of smoothness. This gives, e.g., the estimate

$$\inf_{p\in P_n} \|x^\gamma - p\|_{L_p(I)} \leq C\cdot n^{-2\gamma -2/p} , \quad \gamma > -1/p , \quad n\to\infty ,$$

which is asymptotically optimal (see [11] for p = 2).
So far the approximation theory of the h-p-version is out of the scope of the above approach. This is an open but interesting problem (cf. [1], [19] for some results and motivation).

References

[1] Arnold, D.N., W.L. Wendland, On the asymptotic convergence of collocation methods. Math.Comp. 41 (1983), 349-381.

[2] Aubin, J.-P., Approximation of elliptic boundary value problems. Wiley, N.Y.-London-Sydney-Toronto 1972.

[3] Babuska, I., R.B. Kellogg, J. Pitkäranta, Direct and inverse error estimates for finite elements with mesh refinement. Numer. Math. 33 (1979), 447-471.

[4] Babuska I., W.C. Rheinboldt, A posteriori error analysis of finite element solutions for one-dimensional problems. SIAM J. Numer. Anal. 18 (1981) 3, 565-589.

[5] Bergh J., J. Löfström, Interpolation spaces. An introduction. Springer, Berlin-Heidelberg-New York 1976.

[6] Ciesielski, Z., Equivalence, unconditionality, and convergence a.e. of the spline bases in L_p spaces. Approximation Theory, Banach Center Publ. 4, PWN, Warszawa 1978, 55-68.

[7] DeVore R.A., V. Popov, Interpolation and nonlinear approximation Function spaces and applications. Proc. Lund 1986, Lect. Notes Math. 1302, Springer, Berlin 1988, 191-205.

[8] DeVore, R.A., V. Popov, Interpolation of Besov spaces. Trans. Amer. Math. Soc. 305 (1988) 1, 397-414.

[9] Ditzian, Z., V. Totik, Moduli of smoothness. Springer Series in Comp. Math. 9, Springer, Berlin-Heidelberg-New York 1987.

[10] Elschner, J., G. Schmidt, On spline interpolation in periodic Sobolev spaces. Inst. Math. AdW DDR, Preprint P-Math-01/83, Berlin 1983.

[11] Gui, W., I. Babuska, The h, p and h-p versions of the finite element method in 1 dimension. Numer. Math. 49 (1986) 6, I : 577-612, II : 613-657, III : 659-683.

[12] Nikolskij, S.M., Approximation of functions of several variables and imbedding theorems. 2ed., Moskva, Nauka 1977 (Russian).

[13] Oganesjan, L.A., L.A. Ruhovez, Variational-difference methods for the solution of elliptic equations. Jerevan Univ. Press, Jerevan 1979 (Russian).

[14] Oswald, P., Spline approximation in the metrix L_p, $0<p<1$. Math. Nachrichten 94 (1980), 69-96 (Russian).

[15] Oswald, P., On nonlinear spline approximation in Besov-Sobolev spaces. J. Approx. Theory (to appear).

[16] Oswald, P., On function spaces related to finite element approximation theory. Z. Anal. Anwend. (to appear).

[17] Petrushev, P., Direct and converse theorems for spline and rational approximation and Besov spaces. Function spaces and applications, Proc. Lund 1986, Lect. Notes Math. 1302, Springer, Berlin 1988, 191-205.

[18] Scherer, K., Characterization of generalized Lipschitz classes by best approximation with splines. SIAM J. Numer. Anal. 11 (1974) 2, 283-304.

[19] Scherer, K., On optimal global error bounds obtained by scaled local error estimates. Numer. Math. 36 (1981), 257-277.

[20] Sickel W., Spline representations of functions in Besov-Triebel-Lizorkin spaces (in preparation).

[21] Sjölin, P., J.-O. Strömberg, Basis properties of Hardy spaces Arkiv f. Matematik 21 (1983) 1, 111-125.

[22] Triebel, H., Theory of function spaces. Birkhäuser, Boston 1983 - Akad. Verlagsges. Geest & Portig, Leipzig 1983.

[23] Triebel H., Characterizations of Besov-Hardy-Sobolev spaces: a unified approach. J. Approx. Theory 52 (1988) 2, 162-203.

[24] Triebel, H., Interpolation theory, function spaces, differential operators. North-Holland, Amsterdam-New York-Oxford 1978.

[25] Varga, R.S., Functional analysis and approximation theory in numerical analysis. SIAM, Philadelphia 1971.

[26] Widlund, O., On best error bounds for approximation by piecewise polynomial functions. Numer. Math. 27 (1977), 327-338.

Zur reellen Darstellung periodischer Hermite-Spline-Interpolierender bei äquidistantem Gitter mit Knotenshift.

M. Reimer

1. Einführung

Kürzlich haben Merz und Sippel [5] auch für periodische Hermite-Spline-Interpolierende zu äquidistantem Gitter, allerdings ohne Knotenshift, eine reelle Darstellung angeben können. Die Darstellung wurde unter Fortentwicklung von Ideen in [2], [3], [6] unter zweimaliger Anwendung des Residuensatzes auf komplexem Wege gewonnen. Merz [4] weist bereits darauf hin, daß der in der Zwischenzeit von uns in [7] für den Lagrange-Fall entwickelte reell-algebraische Weg ebenfalls Erfolg versprechend sei. Wir wollen diese Idee hier nicht weiterverfolgen, weil wir inzwischen eine noch sehr viel einfachere und zudem besonders sinnfällige Methode zur reellen Darstellung von periodischen Hermite-Spline-Interpolierenden gefunden haben, bei deren Herleitung nur sehr wenig zu beweisen bleibt. Dabei brauchen wir keinerlei Voraussetzungen bezüglich des Polynomgrades bzw. der Hermite-Ordnung zu machen, wenn wir an den Nicht-Differenzierbarkeitsstellen an Stelle von Ableitungen deren rechtsseitige Grenzwerte interpolieren.

2. Grundlagen

Es seien $m,r,N \in \mathbf{N}$ vorgegeben. Dabei sei $r \leq m$. $S^N_{m,r}$ bezeichne den reellen linearen Raum aller $s \in C^{m-r}(\mathbf{R})$ mit

$$s(j-1+t) = p_j(t)\,, \quad p_j = p_{j+N} \in \mathbf{P}_m, \tag{1}$$

für $0 \leq t \leq 1$, $j \in \mathbf{Z}$. Dabei bezeichne $\mathbf{P}_m$ den Raum aller reellen Polynome vom Grade $\leq m$. Wir setzen (mit einer gegenüber [2]-[8] geänderten Reihenfolge in der Aufführung der Komponenten von p)

$$Z := \begin{bmatrix} 0 & 1 & 0 & \cdots & 0 \\ 0 & 0 & 1 & \cdots & 0 \\ & & \ddots & \ddots & \\ & & & \ddots & \\ & & & \ddots & 1 \\ 1 & \cdots & \cdots & \cdots & 0 \end{bmatrix}_{N \times N}, \quad p := \begin{bmatrix} p_N \\ \cdot \\ \cdot \\ \cdot \\ p_1 \end{bmatrix}$$

für beliebige $p_j = p_{j+N} \in \mathbf{P}_m$, $j \in \mathbf{Z}$, so daß p über (1) genau dann ein $s \in S^N_{m,r}$ charakterisiert, wenn

$$p^{(k)}(0) - Zp^{(k)}(1) = 0 \quad \text{für} \quad k = 0,1,\ldots,m-r \tag{2}$$

gilt (Spline-Bedingungen). Ferner seien ein Shift-Parameter

$$\lambda \in [0,1) \tag{3}$$

und N-periodische Daten

$$y_j^{(k)} = y_{j+N}^{(k)} \in \mathbf{R}\,, \quad j \in \mathbf{Z}\,, \quad k = 0,1,\ldots,r-1,$$

vorgegeben, welche bereits vollständig durch die Vektoren

$$y^{(k)} := (y_N^{(k)},\dots,y_1^{(k)})^T\,, \quad k = 0,1,\dots,r-1, \tag{4}$$

beschrieben werden. Gesucht wird sodann ein $s \in S_{m,r}^N$ mit

$$s^{(k)}(j-\lambda) = y_j^{(k)} \quad \text{für } j \in \mathbb{Z}\,, \quad k = 0,1,\dots,r-1, \tag{5}$$

(Hermite-Interpolations-Aufgabe in $S_{m,r}^N$).

Diese Aufgabe hat im Falle $\lambda = 0$ und $(m+1)/2 < r \leq m$ zunächst keinen Sinn, weil z. B. für $k = r-1$ die Differenzierbarkeitsordnung den durch (2) festgelegten Wert $m-r$ übertrifft. Die Hermite-Interpolations-Aufgabe (5) erhält aber auch in diesem Falle einen Sinn, wenn wir die Interpolations-Bedingungen (5) für $\lambda = 0$ in der Form

$$\lim_{\epsilon \to 0+} s^{(k)}(j+\epsilon) = y_j^{(k)} \tag{5a}$$

für $j \in \mathbb{Z}$, $k = 0,1,\dots,r-1$, interpretieren. Sie ist danach äquivalent mit der Aufgabe, einen Polynomvektor $p \in \mathbf{P}_m^N$ zu finden, der die Spline-Bedingungen (2) sowie die Interpolationsbedingungen

$$p^{(k)}(\tau) = y^{(k)} \quad , \quad k = 0,1,\dots,r-1, \tag{6}$$

für $\tau := 1-\lambda \in (0,1]$ erfüllt.

Wie in [2] - [7] spielen bei uns die verallgemeinerten Euler-Frobenius-Polynome

$$H_n(t,z) := (1-z)^{n+1}\left(t+z\frac{\partial}{\partial z}\right)^n \frac{1}{1-z}, \tag{7}$$

$n \in \mathbb{N}_0$, die entscheidende Rolle. Wir setzen zusätzlich $H_{-n} := 0$ für $n \in \mathbb{N}$. Eine umfassende Darstellung der Eigenschaften der H_n findet man bei Siepmann [8]. Insbesondere gilt mit $.^{(k)} := (\partial/\partial t)^k$

$$H_n'(t,z) = n(1-z)H_{n-1}(t,z) \quad \text{für } n \in \mathbb{Z}, \tag{8}$$

$$H_n^{(k)}(0,z) = z\, H_n^{(k)}(1,z) \quad \text{für } 0 \leq k \leq n-1\,. \tag{9}$$

Wir werden in dieser Arbeit benutzen, daß jede zyklische $N \times N$-Matrix sich in der Form

$$\sum_{j=0}^{N-1} a_j Z^j$$

darstellen läßt und daß $Z^N = I$ gilt. Die Gruppe aller zyklischen $N \times N$-Matrizen über $K \in \{\mathbb{R},\mathbb{C}\}$ ist also isomorph zu $K[z]\,/\,(z^N-1)$, sie ist insbesondere kommutativ. Ferner gilt mit $\zeta := \exp(2\pi i/N)$

$$\operatorname{spec}(Z) = E_N := \{\zeta^k : k = 0,1,\dots,N-1\}\,. \tag{10}$$

Ein Linearfaktor der Gestalt

$$Z - zI$$

ist also genau dann regulär, wenn

$$z \notin E_N$$

gilt. In diesem Fall ist seine Inverse gegeben durch

$$(Z-zI)^{-1} = \frac{1}{1-z^N} \begin{bmatrix} z^{N-1}, z^{N-2}, & \ldots & , 1 \\ 1 \quad , z^{N-1}, z^{N-2}, \ldots, & z \\ \ldots & \\ z^{N-2}, \quad \ldots & , z^{N-1} \end{bmatrix} , \tag{11}$$

sie ist also explizit bekannt. Entsprechend ist jedes Polynom $Q(Z)$ regulär, wenn $Q(z) \neq 0$ für $z \in E_N$ gilt, und man kann $Q(Z)^{-1}$ sehr einfach berechnen, wenn man die Faktorisierung von Q kennt. Schließlich kann man zu gebrochen-rationalen Funktionen

$$R(z) = P(z) \,/\, Q(z)$$

mit Polynomen P und Q, von denen Q auf E_N nicht verschwindet, die Matrix

$$R(Z) := P(Z)Q(Z)^{-1} = Q(Z)^{-1}P(Z)$$

definieren, wobei die Definition von der für R gewählten Darstellung unabhängig ist.

3. Lagrange-Spline-Interpolation (r=1)

Wie bekannt, [8], hat $H_m(\tau,\cdot)$ nur nichtpositive reelle Nullstellen. Wir setzen hier

$$H_m(\tau,-1) \neq 0 \tag{12}$$

voraus, ohne die genauen Bedingungen für diese Situation erneut aufzuzählen. Dann ist $H_m(\tau,Z)$ regulär. Wir setzen

$$p := H_m(\tau,Z)^{-1}H_m(\cdot,Z)y \tag{13}$$

mit $y := y^{(0)}$. Dann gilt wegen $H_m(\cdot,z) \in \mathbb{P}_m$ offensichtlich $p \in \mathbb{P}_m^N$ und die Interpolations—Bedingungen (6) sind von selber erfüllt. Wegen (9) gilt aber zusätzlich

$$H_m^{(k)}(0,Z) = ZH_m^{(k)}(1,Z)$$

für $k = 0,1,\ldots,m-1$. Damit ergibt sich unmittelbar, daß auch die Spline-Bedingungen (2) erfüllt sind. Der Polynomvektor (13) beschreibt bereits die Lösung der Hermite-Interpolations-Aufgabe in $S_{m,1}^N$.

Für die praktische Berechnung von $p(t)$ ist es nützlich, von der Darstellung

$$H_m(t,z) = \sum_{j=0}^{m} a_j(t)\, z^j$$

auszugehen. Die Berechnung von

$$H_m(t,Z)y = \sum_{j=0}^{m} a_j(t)\, Z^j y$$

erfordert dann außer der Berechnung der Koeffizientenpolynome $a_j(t)$ nur die Berechnung einer Linearkomposition der Vektoren $Z^j y$, die ihrerseits "kostenlos" zur Verfügung stehen, wenn man Transfers nicht zählt. Der eigentliche Aufwand bei der Berechnung von $p(t)$ besteht schließlich aus der nachfolgenden Multiplikation mit $H_m(\tau,Z)^{-1}$, welche bis zu m reelle Multiplikationen mit Matrizen der Gestalt (11) erfordert oder mit Hilfe der Schnellen Fourier-Transformation mit $O(N\log N)$ arithmetischen Operationen bewirkt werden kann (Winograd [9]).

Wir wollen noch zwei Querverbindungen herstellen. Die (komplexe) Darstellung von Meinardus und Merz [2] geht aus (13) durch Diagonalisierung hervor.

Um die reelle Darstellung in unserer eigenen Arbeit [6] wiederzugewinnen, jetzt aber ohne den Umweg über die komplexe Darstellung unter Anwendung des Residuenkalküls, gehen wir wie folgt vor: Für $\tau \in (0,1)$ hat $H_m(\tau,\cdot)$ Nullstellen bei $z_1 < z_2 < \dots < z_m < 0$. Wir setzen zusätzlich $z_0 := 1$ und bezeichnen mit $L_0,\dots,L_m \in \mathbf{P}_m$ die Lagrange-Polynome zu diesen Punkten als Knoten. Wegen $H_m(t,1) = m!$ gilt insbesondere

$$H_m(t,z_0)L_0(z) = H_m(\tau,z)\,.$$

Damit erhalten wir entsprechend

$$H_m(t,Z) = H_m(\tau,Z) + \sum_{j=1}^{m} H_m(t,z_j)L_j(Z)\,,$$

und aus (13) folgt

$$p(t) = \{I - \sum_{j=1}^{m} \frac{H_m(t,z_j)}{H_m^z(\tau,z_j)} \cdot C_j(Z)\}\, y\,,$$

wobei die Matrizen

$$C_j(Z) := \frac{1}{1-z_j}(Z-z_jI)^{-1}(Z-I)$$

wegen (11) zur Verfügung stehen. Mit Hilfe der Identität [7, (13)] kann man daraus die

$$p_i(t) = \lambda^{-m}(1-t)^m y_i + \sum_{j=1}^{m} \frac{H_m(t,z_j)}{(1-z_j^N)\, H_m^z(1-\lambda,z_j)} \cdot \sum_{k=1}^{N} z_j^{k-1} y_{i+k}$$

von [7, (22)] wiedergewinnen, die der von uns zuerst in [6] angegebenen entspricht.

4. Hermite-Spline-Interpolation ($r > 1$)

Meinardus und Merz [3] bzw. Merz und Sippel [5] haben die Determinanten

$$\Delta_{m,r}(\tau,z) := \begin{vmatrix} H_m(\tau,z), & H_{m-1}(\tau,z), & \dots, & H_{m-r+1}(\tau,z) \\ H_m'(\tau,z), & H_{m-1}'(\tau,z), & \dots, & H_{m-r+1}'(\tau,z) \\ \cdot & & & \\ \cdot & & & \\ \cdot & & & \\ H_m^{(r-1)}(\tau,z), & H_{m-1}^{(r-1)}(\tau,z), & \dots, & H_{m-r+1}^{(r-1)}(\tau,z) \end{vmatrix}$$

(für $\tau=1$) eingeführt und mit ihrer Hilfe die Lösung der Hermite-Spline-Interpolationsaufgabe unter einigen Zusatzvoraussetzungen konstruiert. Wir folgen zunächst ihrer Grundidee, benutzen dann aber unsere bereits für den Lagrange-Fall entwickelte Methode, was uns nicht nur den Weg durch das Komplexe vermeiden läßt, sondern auch die Betrachtung des allgemeinen Falls ($0 < \tau \leq 1$, $1 \leq r \leq m$, m gerade oder ungerade) gestattet.

Zunächst führen wir zusätzlich die Determinanten

$$\Delta_{m,r,k}(\tau,t,z) := \begin{vmatrix} H_m(\tau,z), & H_{m-1}(\tau,z), & \dots, & H_{m-r+1}(\tau,z) \\ H_m'(\tau,z), & H_{m-1}'(\tau,z), & \dots, & H_{m-r+1}'(\tau,z) \\ \cdots & & & \\ H_m(t,z), & H_{m-1}(t,z), & \dots, & H_{m-r+1}(t,z) \\ \cdots & & & \\ H_m^{(r-1)}(\tau,z), & H_{m-1}^{(r-1)}(\tau,z), & \dots, & H_{m-r+1}^{(r-1)}(\tau,z) \end{vmatrix}$$

für $k = 0,1,\dots,r-1$ ein, die aus $\Delta_{m,r}(\tau,z)$ durch Änderung der $(k+1)$-ten Zeile entstehen. Nach

Konstruktion gilt dann für beliebige τ und z

$$\Delta_{m,r,k}(\tau,\cdot,z) \in \mathbb{P}_m \tag{14}$$

sowie

$$\Delta^{(j)}_{m,r,k}(\tau,\tau,z) = \delta_{jk}\, \Delta_{m,r}(\tau,z) \tag{15}$$

für $j,k \in \{0,1,\dots,r-1\}$. Wir setzen jetzt voraus, daß $\Delta_{m,r}(\tau,z)$ bei festem τ bezüglich der Variablen z nicht identisch verschwindet. Nach Lee und Sharma [1] ist das für $\tau = 1$ und $r \lesssim (m+1)/2$ der Fall. Wir können dann die gebrochen-rationalen Funktionen

$$R_{m,r,k}(\tau,t,z) := \frac{\Delta_{m,r,k}(\tau,t,z)}{\Delta_{m,r}(\tau,z)} \tag{16}$$

für $k = 0,1,\dots,r-1$ einführen. Wegen (15) gilt dann

$$R^{(j)}_{m,r,k}(\tau,\tau,z) = \delta_{jk} \tag{17}$$

für $j,k \in \{0,1,\dots,r-1\}$ und alle z. Wegen (8) und (9) gilt ferner

$$R^{(j)}_{m,r,k}(\tau,0,z) = z\, R^{(j)}_{m,r,k}(\tau,1,z) \tag{18}$$

für $k \in \{0,1,\dots,r-1\}$ und $j \in \{0,1,\dots,m-r\}$.

Da $\Delta_{m,r}(\tau,z)$ nicht das Nullpolynom ist, existiert der größte gemeinsame Teiler der Polynome

$$\Delta_{m,r}(\tau,z) \text{ und } \Delta_{m,r,k}(\tau,t,z)\,, \quad k \in \{0,1,\dots,r-1\}$$

bezüglich der Variablen t und z. Man beachte dabei, daß das erste Polynom nur von einer Variablen abhängt. Wir bezeichnen ihn mit $D_{m,r}(\tau,z)$ und setzen

$$Q_{m,r}(\tau,z) := \Delta_{m,r}(\tau,z)/D_{m,r}(\tau,z)\,,$$

$$P_{m,r,k}(\tau,t,z) := \Delta_{m,r,k}(\tau,t,z)/D_{m,r}(\tau,z)$$

für $k = 0,1,\dots,r-1$, so daß auch

$$R_{m,r,k}(\tau,t,z) = P_{m,r,k}(\tau,t,z)/Q_{m,r}(\tau,z)$$

für $k = 0,1,\dots,r-1$ gilt. Danach machen wir die Voraussetzung

$$Q_{m,r}(\tau,z) \neq 0 \quad \text{für} \quad z \in E_N\,, \tag{19}$$

so daß $Q_{m,r}(\tau,Z)$ regulär und

$$R_{m,r,k}(\tau,t,Z) = Q_{m,r}(\tau,Z)^{-1}\, P_{m,r,k}(\tau,t,Z)$$

für $k = 0,1,\dots,r-1$ und $t \in \mathbb{R}$ erklärt ist. Wegen (17) und (18) gilt dabei

$$R^{(j)}_{m,r,k}(\tau,\tau,Z) = \delta_{jk}\, I \tag{20}$$

für $j,k \in \{0,1,\dots,r-1\}$ und

$$R^{(j)}_{m,r,k}(\tau,0,Z) = Z\, R^{(j)}_{m,r,k}(\tau,1,Z) \tag{21}$$

für $k \in \{0,1,\dots,r-1\}$, $j \in \{0,1,\dots,m-r\}$.

Damit sind wir am Ziel. Wir setzen

$$p(t) := \sum_{k=0}^{r-1} R_{m,r,k}(\tau,t,Z)y^{(k)}. \tag{22}$$

Dann ist $p \in \mathbb{P}_m^N$ nach Konstruktion. Wegen (21) sind die Spline-Bedingungen (2), wegen (20) die Interpolationsbedingungen (6) erfüllt. Also charakterisiert p eine Lösung s der Hermite–Interpolations-Aufgabe in $S_{m,r}^N$. Sie ist eindeutig bestimmt, da die Lösbarkeit für beliebige Interpolationsdaten gesichert ist.

Bei der praktischen Berechnung von $p(t)$ ist es günstiger, von der mit (22) äquivalenten Darstellung

$$p(t) = Q_{m,r}(\tau,Z)^{-1} \sum_{k=0}^{r-1} P_{m,r,k}(\tau,t,Z)y^{(k)} \tag{23}$$

auszugehen.

5. Quadratische Hermite-Spline-Interpolation, (m,r) = (2,2)

Für $\tau = 1$ $(\lambda = 0)$ sind in diesem Falle unter anderem die Werte $y'(j+0)$ zu interpolieren. Nach Reimer und Siepmann [7] gilt für $n \in \mathbb{N}_0$

$$H_n(t,z) = \sum_{j=0}^{n} c_j^n(t)\, z^j(1-z)^{n-j}$$

mit

$$c_j^n(t) = \Delta^j(t+\cdot)^n$$

für $j = 0,1,\ldots,n$. Man findet damit

$$\begin{aligned} H_0(t,z) &= 1, \\ H_1(t,z) &= t(1-z) + z, \\ H_2(t,z) &= t^2(1-z)^2 + (2t+1)z(1-z) + 2z^2, \\ H_3(t,z) &= t^3(1-z)^3 + (3t^2+3t+1)z(1-z)^2 + 6(t+1)z^2(1-z) + 6z^3. \end{aligned}$$

Damit ergibt sich

$$\Delta_{2,2,0}(\tau,t,z) = (1-z)^2\{(1-t^2-2\tau(1-t))z + t(t-2\tau)\} \quad ,$$

$$\Delta_{2,2,1}(\tau,t,z) = (t-\tau)(1-z)^2\{(1-t)(1-\tau)z - t\tau\}, \tag{24}$$
$$\Delta_{2,2}(\tau,z) = \Delta_{2,2,0}(\tau,\tau,z) = (1-z)^2\{(1-\tau)^2 z - \tau^2\}.$$

Die Verhältnisse werden im Falle $\tau = 1$ besonders übersichtlich, da $\Delta_{2,2}$ die Polynome $\Delta_{2,2,0}$ und $\Delta_{2,2,1}$ teilt. Wir erhalten in diesem Falle die Lösung der Hermite-Interpolations-Aufgabe in $S^N_{2,2}$ in der Form

$$p(t) = [(t-1)^2 Z - t(t-2)I]y^{(0)} + t(t-1)Iy^{(1)}. \tag{25}$$

Die Berechnung von $p(t)$ kann mit nur $3N+2$ Multiplikationen und $2N+2$ Additionen bewirkt werden.

Im weiteren sei $\tau \in (0,1)$. Dann ist der größte gemeinsame Teiler der Polynome (24) gegeben durch

$$D_{2,2}(\tau,z) = (1-z)^2,$$

und wir erhalten

$$Q_{2,2}(\tau,z) = (1-\tau)^2 z - \tau^2$$

Die Voraussetzung (19) ist genau dann erfüllt, wenn $\tau \neq \frac{1}{2}$ ist. Für $\tau \in (0,1) \setminus \{\frac{1}{2}\}$ hingegen erhalten wir die eindeutige Lösung unseres Problems in der Form

$$p(t) = \frac{1}{(1-\tau)^2} \left(Z - \left(\frac{\tau}{1-\tau}\right)^2 I\right)^{-1}$$
$$\cdot \{[(1-t^2-2\tau(1-t))\, Z + t(t-2\tau)I]\, y^{(0)}$$
$$+ (t-\tau)[(1-\tau)(1-t)\, Z - t\tau I]\, y^{(1)}\}$$

Man beachte wieder, daß die Berechnung der Vektoren $Z^j y^{(k)}$, $j,k \in \{0,1\}$ entfällt, da sie durch Transfers ersetzt werden kann, und daß die Matrix $(Z - (\frac{\tau}{1-\tau})^2 I)^{-1}$ explizit bekannt ist.

6. Kubische Hermite-Spline-Interpolation, (m,r) = (3,2)

Hier ist $r = (m+1)/2$, es kommen also bei der Interpolation nur Ableitungen vor. Man findet nach einiger Rechnung:

$$\Delta_{3,2,0} = (1-z)^3 \; \{ (1-\tau)(1-t)[2t^2 - (3\tau+1)t + 3\tau-1] \, z^2 + [3\tau^2(2t^2-2t-1) - 2\tau(2t^3-3t-1) + 2t^3-3t^2] \, z + [2\tau t^3 - 3\tau^2 t^2] \} ,$$

$$\Delta_{3,2,1} = - (t-\tau)(1-z)^3 \{ (1-\tau)^2(1-t)^2 z^2 + [t^2+\tau^2-t-\tau-2\tau(1-\tau)t(1-t)] \, z + \tau^2 t^2 \} ,$$

$$\Delta_{3,2}(\tau,z) = \Delta_{3,2,0}(\tau,\tau,z) = - (1-z)^3 \{ (1-\tau)^4 z^2 - 2\tau(1-\tau)(1+\tau-\tau^2) \, z + \tau^4 \} .$$

Für $\tau = 1$ teilt $\Delta_{3,2}$ wieder die Polynome $\Delta_{3,2,0}$ und $\Delta_{3,2,1}$, und wir erhalten die Lösungsformel

$$p(t) = [(t-1)(2t^2-t-1) \; Z \; + \; t^2(3-2t) \; I] \; y^{(0)} + (t-1) \; [t(t-1) \; Z \; + t^2 I] \; y^{(1)}$$

Für $\tau \in (0,1)$ hat $\Delta_{3,2}(\tau,z)$ bei $z = 1$ eine dreifache Nullstelle. Zusätzlich hat $\Delta_{3,2}(\tau,z)$ zwei Nullstellen bei

$$z_{1,2} = \frac{\tau}{(1-\tau)^3} \left\{ 1 + \tau - \tau^2 \pm \sqrt{1 + 2\tau - 2\tau^2} \right\} .$$

Diese sind positiv und voneinander verschieden. Genau dann, wenn

$$\tau \in \left\{ \tfrac{1}{2}(1 + \sqrt{\tfrac{1}{3}}) \; , \; \tfrac{1}{2}(1 - \sqrt{\tfrac{1}{3}}) \right\}$$

gilt, hat eine dieser beiden Nullstellen den Wert 1. Gleichzeitig hat $\Delta_{3,2,1}(\tau,1,z)$ die Nullstelle $z = 1$ nur von der Vielfachheit 3. Also hat $R_{3,2,1}(\tau,1,z)$ bei $z = 1$ eine nicht-hebbare Singularität. Unter der Voraussetzung

$$\tau \in (0,1) \setminus \left\{ \tfrac{1}{2}(1 + \sqrt{\tfrac{1}{3}}), \; \tfrac{1}{2}(1 - \sqrt{\tfrac{1}{3}}) \right\}$$

ist die Hermite-Interpolationsaufgabe in $S^N_{3,2}$ allerdings wieder eindeutig lösbar.
Die Lösung kann mit Hilfe der Polynome

$$P_{3,2,0}(\tau,t,z) = \Delta_{3,2,0}(\tau,t,z) \; / \; (1-z)^3 ,$$

$$P_{3,2,1}(\tau,t,z) = \Delta_{3,2,1}(\tau,t,z) \,/\, (1-z)^3,$$
$$Q_{3,2}(\tau,z) \quad = (1-\tau)^4 z^2 - 2\tau(1-\tau)(1+\tau-\tau^2)\, z + \tau^4$$

wie in (23) angegeben explizit beschrieben werden, da die Voraussetzung (19) wieder erfüllt ist.

7. Schlußbemerkung

Nach Lee und Sharma [1] entarten die Polynome $\Delta_{m,r}(\tau,z)$ für $\tau = 1$ dem Grade nach. Das haben wir an den Beispielen von Abschnitt 5 und Abschnitt 6 auch bestätigt gefunden. Dies gilt jedoch nicht mehr im Falle $\tau \neq 1$, was sich auch im Auftreten weiterer Nullstellen bemerkbar macht.

Literaturverzeichnis

[1] S. L. Lee, A. Sharma: Cardinal lacunary interpolation by g-splines I. The characteristic polynomials, J. Approx. Th. 16 (1976) 85 – 96.

[2] G. Meinardus, G. Merz: Zur periodischen Spline-Interpolation. In: Spline-Funktionen (K. Böhmer, G. Meinardus, W. Schempp, eds.), pp. 177 – 195. Mannheim: Bibliographisches Institut 1974.

[3] G. Meinardus, G. Merz: Hermite-Interpolation mit periodischen Spline-Funktionen. Numerical Methods of Approximation Theory, ISNM 52 (1980) 200 – 210.

[4] G. Merz: The fundamental splines of periodic Hermite interpolation for equidistant lattics, Kasseler Schriften zur Angewandten Mathematik Nr. 1, 1986.

[5] G. Merz, W. Sippel: Zur Konstruktion periodischer Hermite-Interpolationssplines bei äquidistanter Knotenverteilung, J. Approx. Th. 54, (1988).

[6] M. Reimer: Extremal spline bases, J. Approx. Th. 36 (1982) 91 – 98.

[7] M. Reimer, D. Siepmann: An elementary algebraic representation of polynomial spline interpolants for equidistant lattics and its condition, Numer. Math. 49 (1986) 55 – 65.

[8] D. Siepmann: Kardinale Spline-Interpolation bezüglich äquidistant verteilter Knoten, Diss. Dortmund 1984.

[9] S. Winograd: On the number of multiplications required to compute certain functions, Proc. N. A. S. 58 (1967) 1840 – 1842.

Adresse:

Prof. Dr. M. Reimer
Universität Dortmund
Lehrstuhl Mathematik III
Postfach 50 05 00
D-4600 Dortmund 50
Bundesrepublik Deutschland

SAFE ERROR BOUNDS FOR THE DISCRETIZATION OF ELLIPTIC BOUNDARY VALUE PROBLEMS USING ENCLOSURES OF THE SOLUTION

Hans-Görg Roos [1)]

1. Introduction

Despite the significant advances made on the theory and implementation of discretization methods for elliptic boundary value problems, the selection of a discretization technique for a particular problem is largely based on intuition and experience obtained from solving similar tasks. Moreover, the assessment of the reliability of the discrete solution continues to be the most difficult part of the numerical analysis, safe error estimation and control becomes now the most interesting aspect of the discretization technology.

The discretization error represents the difference between the exact and the discrete solution for the given mathematical model. In general, the exact solution is not available but nevertheless it is possible to construct quantitative estimates for the discretization error in some cases. For instance, for linear elliptic second order boundary value problems using linear finite elements Lehmann [13] proved

$$\|u-u_h\|_1 \le 0.8123\, C\sqrt{\max_{T\in T_h} C(T)}\; |u|_2 \,, \qquad (1)$$

where C and C(T) (T is a single triangle) are computable constants. A final problem for obtaining computable a priori bounds consists in estimating the seminorm of the exact solution in (1).

Such a priori bounds yield in general no information about the local behaviour of the error with respect to an actual finite element model. Thus, such bounds are less useful for the adaptive improvement of the discrete solution.

A posteriori estimates use information obtained during the solution process. In recent years considerable effort has been devoted to the development of a posteriori error estimates based on local error estimators. Local error estimators allow an adaptive grid generation, the computed local information is used for steering the process.

One way to construct local error estimators is based on a modification of the error bound (1). Setting

$$\eta(T) = \sqrt{C(T)}\; |u_h|^*_{2,T} \qquad (2)$$

1) Technical University of Dresden, Department of Mathematics
8027 Dresden, Mommsenstr. 13, German Democratic Republic

we obtain a local error estimator supposed $|u_h|^*_{2,T}$ approximates $|u|_{2,T}$ (u_h denotes the actual finite element solution). Details and other typs of local error estimators are discussed in [4],[8],[9].

There are no local error estimators known which yield safe error bounds In contrast to the local error estimator strategy it is conceivable to control the error using upper and lower solutions of the original problem and to steer a grid using information on the difference of the upper and lower solution [10]. In this paper we discuss the actual situation concerning the construction of upper and lower solutions for elliptic boundary value problems in a numerical way.

2. A critical view on some enclosing discretization techniques

Let us consider the semilinear boundary value problem

$$\begin{aligned} Lu &= 0 \quad \text{in } \Omega \subset R^2 \\ u &= 0 \quad \text{on } \partial\Omega = \Gamma \end{aligned} \tag{3}$$

with the uniformly elliptic operator L

$$Lu := -\sum_{i,j} a_{ij} \frac{\partial^2 u}{\partial x_i \partial x_j} + \sum_i b_i \frac{\partial u}{\partial x_i} + g(x,u).$$

We assume $\frac{\partial g}{\partial u} \geq 0$, further a_{ij}, b_i to be sufficiently smooth.

Classical bounds for the solution of (3) are based on the strong maximum principle and work with $C^2(\bar{\Omega})$-class functions.
Such functions allow only rough bounds, $C^0(\bar{\Omega})$-or $C^1(\bar{\Omega})$-splines which are typical for finite element or collocation methods do not belong to this class.

Werner [17] was the first who tried to construct upper and lower bounds for linear elliptic problems (i.e. $g(x,u) := c(x)u - f(x)$, $c \geq 0$) using finite elements. At first he proved a generalized comparison theorem. Let us assume that Ω is subdivided in a finite number of domains Ω_i with boundary Γ_i, assume $v_1, v_2 \in C(\bar{\Omega})$ with $v_1, v_2|_{\bar{\Omega}_i} \in C^1(\bar{\Omega}_i) \cap C^2(\Omega_i)$. Then from

(C_1) $Lv_1 \geq Lv_2$ in Ω_i

(C_2) $v_1 \geq v_2$ on $\partial\Omega$

(C_3) $Sv_1 \geq Sv_2$ on Λ $(\Lambda = \{x \in \cup\Gamma_i,\ x \notin \partial\Omega)$

it follows $v_1(x) \geq v_2(x)$ for all $x \in \bar{\Omega}$.

S characterizes some kind of jump of the derivative crossing from one subdomain to some other (compare [17]). For us it is essential to note that condition (C_3) is fulfilled if $v_{1,2} \in C^1(\bar{\Omega})$.

Now let us assume that u_h belongs to some spline space. u_h^+ is an upper solution (a lower solution u_h^- is constructed in an analogous way) if

$$Lu_h^+ \geq 0 \text{ in } \Omega_i, \; u_h^+ \geq 0 \text{ on } \partial\Omega \text{ and } Su_h^+ \geq 0 \text{ on } \Lambda. \tag{4}$$

Werner [17] and in a larger extent Meyn [14] succeeded in formulating a linear set of inequalities LIN which is sufficient for (4), at least for special differential operators and special spline functions. To get bounds in a given point $x \in \Omega$ the following linear optimization problem has to be solved:

$$u_h^+(x) - u_h^-(x) \to \min \text{ under the restrictions LIN.} \tag{5}$$

The essential disadvantage of this method consists in the enormous expense of the method - in every point one has to solve a linear optimization problem. Of course, it would be preferable to compute the unknown coefficients in some spline function $u_h^+(x)$ in such a way that $u_h^+(x) \geq u(x)$ for all $x \in \bar{\Omega}$.

Let us denote by P_k the space of piecewise polynomials of degree not greater than k and let us choose $u_h \in C^1(\bar{\Omega}) \cap P_k$. Under the very restrictive hypothesis $u \in C^{k+1}(\bar{\Omega})$ Meyn [14] was able to prove

$$u_h^+ - u_h^- = O(h^{k-1}). \tag{6}$$

It is well known that the principle of monotone iteration permits the generation of monotone enclosing sequences. If we denote by L^* some linear part of L

$$L^* := -\sum_{i,j} a_{ij} \frac{\partial^2}{\partial x_i \partial x_j} + \sum_i b_i \frac{\partial}{\partial x_i}$$

then the original version of monotone iteration is characterized by

$$(L^* + \rho I)u^{m+1} = (\rho I - G)u^m \qquad (m=0,1,\ldots) \tag{7}$$

(G corresponds to the nonlinearity, ρ is some suitable chosen constant I the identical mapping). Choosing the initial values to be upper and lower solutions, respectively, one obtains

$$\underline{u}^0 \leq \underline{u}^1 \leq \ldots \leq \underline{u}^m \leq \underline{u}^{m+1} \leq \ldots \leq u \leq \ldots \leq \bar{u}^{m+1} \leq \bar{u}^m \leq \ldots \leq \bar{u}^1 \leq \bar{u}^0.$$

Unfortunately, the iteration process (7) is not numerically constructive because in every iteration step one has to solve a linear elliptic boundary value problem exactly.

Recently Großmann [3] proposed the so-called MID-technique (monotone iteration discretization). In this technique we choose a finite dimensional space W_h (for instance, $W_h = P_k$), ρ piecewise constant and some auxiliary operator Q. The iteration process (7) is modified to

$$(L^* + \rho I)u^{m+1} = s^m; \; s^{m+1} = \min(s^m, Qu^{m+1}) \tag{8}$$

with $s^m \in W_h$ (Q is a mapping in W_h, too). The finite dimensional performance of the MID-technique rests on patched base functions. u^m is an element of the space $(L^*+\rho I)^{-1}W_h$. Choosing a base in W_h one obtains theoretically a base in $(L^*+\rho I)^{-1}W_h$ and thus a finite dimensional representation of u^m. Practically this fact seriously restricts the class of differential operators which can be treated to one dimensional operators with piecewise constant (in some extent with piecewise polynomial) coefficients. Some new considerations to approximate bases [1] do not change the situation basically.

Therefore we discuss in chapter 3 a new rigorous approach to construct upper and lower solutions based on global defect estimations.

3. C^1-methods for second order linear boundary value problems combined with a shift of the defect

In this chapter let us consider the linear boundary value problem

$$Lu := -\Delta u + b\cdot\nabla u + cu = g \text{ in } \Omega \subset R^2 \qquad (9)$$

$$u = 0 \text{ on } \Gamma,$$

assuming $g \in C(\bar{\Omega})$, $c - \operatorname{div} b \geq \beta > 0$. The corresponding weak formulation is characterized by: Find some $u \in H_0^1(\Omega)$ such that

$$a(u,v) = (g,v) \text{ for all } v \in H_0^1(\Omega), \qquad (10)$$

where $a(u,v) := (\nabla u,\nabla v)+(b\cdot\nabla u,v)+(cu,v)$ denoting by (,) the scalar product in the space $L^2(\Omega)$.

We discuss three discretization methods - finite elements, collocation and an adjoint H^{-1}-Galerkin technique. Let us denote by V_h some finite dimensional subspace of $V = H_0^1(\Omega)$. While in the finite element method it is in general use to choose spline functions $v_h \in V_h \subset C^0(\bar{\Omega})$ for second order problems the collocation method and the H^{-1}- Galerkin technique require $V_h \subset C^1(\Omega)$. Because for our purposes C^1-elements are advantagous we concentrate ourselve on the case $V_h \subset C^1(\Omega)$.

Let us assume that $v_h \in V_h$ is piecewise polynomial (polynomial on every subdomain Ω_i). Then, corresponding to the above cited comparison theorem of Werner, from

$$\begin{aligned} Lv_h - g &\geq 0 \text{ on } \Omega_i \text{ for all subdomains} \\ v_h &\geq 0 \text{ on } \Gamma \end{aligned} \qquad (11)$$

it follows $v_h(x) \geq u(x)$ for all $x \in \bar{\Omega}$:

In the finite element method $u_h \in V_h$ is characterized by

$$a(u_h,w_h) = (g,w_h) \text{ for all } w_h \in V_h.$$

In the collocation method one has to define collocation points ξ_ℓ ($\ell=1,\dots,N$) with $N = \dim V_h$, $\xi_\ell \in \operatorname{int} \Omega_i$. The approximate solution satisfies

$$Lu_h(\xi_\ell) = g(\xi_\ell) \quad (\ell=1,\dots,N).$$

The adjoint H^{-1}-Galerkin technique is described by

$$(Lu_h, w_h^*) = (g, w_h^*) \text{ for all } w_h^* \in W_h$$

where $W_h \subset L^2(\Omega)$ is some test space with $\dim W_h = \dim V_h$.

Now we denote by u_h some approximate solution of the original problem (9) computed using finite elements, collocation or adjoint H^{-1}-Galerkin. In the next step we calculate the local defect $\sup_{x\in\Omega_i} |Lu_h(x)-g(x)|$ on every element and finally the global defect ε - the L^∞-norm $\|Lu_h-g\|_\infty$.

Then we realize a shift of the defect using a majorizing element z with respect to the given differential operator. Such an element satisfies $z \in C^2(\Omega) \cap C(\bar{\Omega})$ and

$$Lz \geq 1, \; z|_\Gamma \geq 0 \; .$$

From the theory of singular perturbations of elliptic boundary value problems it is wellknown that the construction of majorizing elements for wide classes of operators is possible. For instance, in the case $L = -\Delta$, $\Omega \subset [0,a] \times [0,b]$ we choose

$$z = \frac{1}{4} (x(a-x) + y(b-y)).$$

Now, let us set

$$u_h^* = u_h + \varepsilon z. \tag{12}$$

Evidently it holds $Lu_h^* = Lu_h + \varepsilon Lz \geq Lu_h + \varepsilon \geq g$.

<u>Theorem 1</u>: Assume $u_h \in C^1$ to be some approximate solution of the given problem (1). Then, the corresponding to (12) shifted function u_h^* represents an upper solution of the original problem, $u_h^*(x) \geq u(x)$ for all $x \in \bar{\Omega}$.

It is possible to construct some upper solution $\tilde{u}_h \in V_h$, too. Let us denote by π the interpolation operator into V_h. Setting

$$\tilde{u}_h = u_h + 2\varepsilon\pi z$$

we obtain

$$L\tilde{u}_h = Lu_h + 2\varepsilon Lz + 2\varepsilon L(\pi z - z) \geq Lu_h + \varepsilon$$

provided $\|L(\pi z-z)\|_\infty \leq \frac{1}{2}$ which is satisfied for $h \leq h_o$, h_o sufficiently small (V_h is assumed to be some standard finite element space).

Now, let us generally suppose that the decomposition of $\bar{\Omega}$ corresponds to an admissible, regular decomposition in the context of finite elements and that V_h is some C^1 finite element space. Our regularity assumption safeguards especially the validity of the inverse inequalities

$$\|v_h\|_\infty \le \frac{C}{h}\|v_h\|_0,\ \|v_h\|_2 \le \frac{C}{h}\|v_h\|_1 \quad (v_h \in V_h) \tag{13}$$

$$\|v_h\|_{1,\infty} \le \frac{C}{h}\|v_h\|_\infty \quad \text{(C does not depend on h).}$$

Next we investigate for some special discretization methods whether or not the upper solution u_h^* converges to the exact solution for $h \to 0$. The triangle inequality results in

$$\|u-u_h^*\|_\infty \le \|u-u_h\|_\infty + C\|Lu_h-g\|_\infty \le C\max(\|u-u_h\|_\infty, \|u-u_h\|_{1,\infty}, \|u-u_h\|_{2,\infty}). \tag{14}$$

Thus we expect that the order of convergence of our upper solution to the exact solution meets at least the convergence order of u_h in the $H^{2,\infty}$-norm.

Example 1: one-dimensional C^1-collocation

Let us choose $u_h \in C^1 \cap P_k$ $(k \ge 2)$ and k-1 collocation points $\xi_{i,\ell}$ $(\ell=1,\ldots,k-1)$ in every subinterval (x_i, x_{i+1}) with $\xi_{i,\ell} = \frac{x_{i+1}+x_i}{2} + \rho_\ell \frac{x_{i+1}-x_i}{2}$.

The parameters ρ_ℓ $(-1 < \rho_\ell < 1)$ are stated in such a way that

$$\int_{-1}^{1} q \prod_{\ell=1}^{k-1} (x-\rho_\ell)\,dx = 0 \text{ for all } q \in P_{s-1}.$$

It is well known that the maximal admissible value s is equal to $s = k-1$, the corresponding collocation points are the Gauss-Legendre points. In [7] we find the following convergence result:

$$\|u-u_h\|_{i,\infty} \le C\, h^{k-1+\min(s,2-i)} \quad (i=0,1,2). \tag{15}$$

Superconvergence results in the grid points especially for Gauss-Legendre collocation are less interesting for our purposes. Combining (14) and (15) it seems that the choice of the collocation points does not influence on the convergence order of our upper solution.

Proposition 1: $C^1 \cap P_k$-collocation in the one-dimensional case results for arbitrary choosen collocation points in combination with a global shift of the defect in an $O(h^{k-1})$ upper solution error.

In [11] we described an enclosing discretization technique of arbitrary order for solving the boundary value problem

$$-u'' = g(x,u),\ u(0) = u(1) = 0\ .$$

The basic idea of this method consists in replacing g by a piecewise polynomial $\bar{g}$ of degree k-2 $(k \ge 2)$ such that $\bar{g} \ge g$. u_h is defined to be the exact solution of the boundary value problem

$$-u_h'' = \bar{g}(x,u_h),\ u_h(0) = u_h(1) = 0\ .$$

Thus, $u_h \in C^1 \cap P_k$ holds and in some cases with respect to the choice of $\bar{g}$ it is possible to interprete this technique again as a collocation technique [12]. The convergence order k-1 meets the convergence order of the global defect correction technique described above.

Example 2: two-dimensional C^1-collocation on a rectangular grid

Let us consider the linear elliptic boundary value problem (9) in a rectangular domain, for instance in $\Omega = (0,1)^2$. We decompose Ω in rectangles and choose $u_h \in (C^1 \cap P_k)^2$ with $k \geq 3$. In every subrectangle we define the collocation points ($(k-1)^2$) to be the corresponding Gauss-Legendre points.

Following [15] we have for $u \in H^{k+3}(\Omega)$ the error estimate

$$\|u-u_h\|_m = O(h^{k+1-m}) \quad (m=0,1). \tag{15}$$

Again we denote by $\pi u \in V_h = (C^1 \cap P_k)^2$ the V_h-interpolate of u. The triangle inequality

$$\|u-u_h\|_2 \leq \|u-\pi u\|_2 + \|\pi u-u_h\|_2, \tag{16}$$

an inverse inequality

$$\|u-u_h\|_2 \leq \|u-\pi u\|_2 + \frac{C}{h}\|\pi u-u_h\|_1 \leq \|u-\pi u\|_2 + \frac{C}{h}(\|u-\pi u\|_1 + \|u-u_h\|_1) \tag{17}$$

and standard interpolation results yield

$$\|u-u_h\|_2 = O(h^{k-1}). \tag{18}$$

The application of an inverse inequality leads to

Proposition 2: $(C^1 \cap P_k)^2$ - Gauss-Legendre collocation ($k \geq 3$) in combination with a global shift of the defect results in an upper solution which tends to the exact solution for $h \to 0$. The convergence rate in the L^∞-norm is at least k-2.

The comparison of proposition 2 with the one-dimensional case shows that we losed one power of h. This effect likely is due to our rough way in deriving $H^{2,\infty}$-bounds from H^2-bounds using an inverse inequality.

Remark: Choosing $W_h = (P_{k-2} \cap C_{-1})^2$ in the adjoint H^{-1}-Galerkin technique (C_{-1} characterizes that $w_h \in W_h$ is not necessary continuous) proposition 2 remains true because the asymptotic estimate (15) holds for the H^{-1}-technique, too.

Example 3: C^1-finite elements in two dimensions

Let us consider some polygonal domain Ω and an admissible, regular triangulation of the domain. It is well known that finite element spaces which base on real polynomials on every triangle are very complicated, for instance the Argyris or the Bell triangle. In the last years some elements which use piecewise polynomials on every triangle of the given triangulation became very popular. It is fashionable to use piecewise quadratic (Powell-Sabin) or piecewise cubic (Clough-Tocher) polynomials, in every case therefore it is necessary to subdivide every triangle in some auxiliary triangles.

First we consider the Clough-Tocher element with 12 degrees of freedom. $v_h \in V_h \subset C^1$ is piecewise cubic on every of the three subtriangles of the basic triangles. Due to [6] we have the following interpolation results:

$$\|v-\pi v\|_m \leq C\, h^{4-m} |v|_4 \text{ for } v \in H^4(\Omega) \text{ and } m=0,1,2. \tag{19}$$

Cea's lemma and this interpolation result immediately yield the H^1-error bound

$$\|u-u_h\|_1 = O(h^3)$$

provided $u \in H^4(\Omega)$. Again we use (16),(17) to deduce H^2- and $H^{2,\infty}$-bounds and finally to estimate $\|u-u_h^*\|$ based on the inequality (14).

Proposition 3: The discretization of the elliptic boundary value problem (9) using the Clough-Tocher element in combination with a global shift of the defect results in an upper solution which tends to the exact solution for $h \to 0$. The convergence rate in the L^∞-norm is at least one.

For the Powell-Sabin element with 9 degrees of freedom and a subdivision of every basic triangle in 6 auxiliary triangles Buhl [5] announced the interpolation result

$$\|v-\pi v\|_m \leq C\, h^{3-m} |v|_3 \text{ for } v \in H^3(\Omega) \text{ and } m=0,1,2. \tag{20}$$

As a direct consequence we obtain

$$\|u-u_h\|_1 = O(h^2) \quad \text{for } u \in H^3(\Omega),$$

the usual convergence rate for the discretization of second order problems based on quadratic elements. We expect the convergence of u_h^* to the exact solution in this case, too. But, of course, we have to refine our convergence analysis.

Finally let us remark that it would be possible to analyze rectangular C^1-finite elements in a similar way.

References

[1] Abhari,H., Großmann,Ch.: Approximate bases for boundary value problems and applications to MID-techniques. Preprint 07-09-88, TUD 1988.

[2] Adams,E., Ansorge,R., Großmann,Ch., Roos,H.-G. (eds.): Discretization in differential equations and enclosures. Akademie-Verlag, Berlin 1987.

[3] Al-Zanaidi,M., Großmann,Ch.: Monotone iteration discretization algorithm for BVP's. (to appear in Computing).

[4] Babuska,I., Zienkiewicz,O.C., Gago,J., Oliviera,E.R. de A.: Accuracy estimates and adaptive refinements in finite element computations. John Wiley, New York 1986.

[5] Buhl,H.J.: Stückweise quadratische C^1-Interpolation und ihr Einsatz bei der Galerkindiskretisierung im Raume $H^2(\Omega)$. Dissertation, Wuppertal 1987.

[6] Ciarlet,P.: Finite element methods for elliptic equations. Amsterdam 1977.

[7] de Boor,C.: A practical guide to splines. Berlin, Heidelberg, New York 1978.

[8] Demkowicz,L., Devloo,Ph., Oden,J.T.: On an h-type mesh-refinement strategy based on minimization of interpolation errors. Computer Math. in Appl.Mech.Eng., 53(1985), 67-89.

[9] Eriksson,K., Johnson, Cl.: An adaptive finite element method for linear elliptic problems. Math.Comput., 50(1988), 182, 361-383.

[10] Großmann,Ch., Roos,H.-G.: Feedback grid generation via monotone discretization for two-point boundary value problems. IMA J.Numer.Anal., 6(1986), 421-432.

[11] Großmann,Ch., Roos,H.-G.: Uniformly enclosing discretization methods of high order for boundary value problems (to appear in Math. of Comput.).

[12] Kadner,H., Roos,H.-G.: Die Anwendung von Kollokationsverfahren zur effektiven Lösung von Randwertproblemen. Wiss.Z. TUD, 37(1988), 171-173.

[13] Lehmann,R.: Computable error bounds in the finite element method. IMA J.Numer.Anal. 6(1986).

[14] Meyn,K.H.: Monotonieaussagen für elliptische und parabolische Randwertaufgaben und Anwendung auf Finite-Elemente-Funktionen. Dissertation, Universität Hamburg 1979.

[15] Percell,P., Wheeler,M.F.: A C^1-finite element collocation method for elliptic equations. SIAM J.Num.Anal., 17(1980), 605-622.

[16] Roos,H.-G.: Uniformly enclosing discretization methods for semilinear boundary value problems. (to appear in Banach Center Publications).

[17] Werner,B.: Monotonie und finite Elemente bei elliptischen Differentialgleichungen. ISNM 27(1975), 309-329.

QUADRATIC SPLINE INTERPOLATION ON UNIFORM MESHES

S. Sallam and M.N. El-Tarazi [1)]

ABSTRACT.

The interpolation of a discrete set of data on a uniform mesh, representing the derivative values of a smooth function f between consecutive nodes, is obtained using quadratic splines. An L_∞-error estimate for the constructed spline is also presented.

AMS Subject Classification: 65D05, 41A10

Key words: Interpolation, quadratic spline,

1. INTRODUCTION.

The basic aim of this paper is to construct and study a quadratic interpolatory spline that matches the derivative values of a smooth function f at arbitrary positions [dependant on a single parameter $\lambda(0\leq\lambda\leq 1)$] between consecutive knots, and matches the function values at the end points of the domain of f. In Section 2, existence and uniqueness of such a spline is proved for uniform distribution. An investigation of the resulting error in the uniform norm is presented in Section 3, where $f \in C^{\ell}[0,1]$, $\ell=2,3$. Section 4 will be devoted to a complete study of the special cases $\lambda = 0$ and $\lambda = 1$ to indicate that the error bounds obtained in Section 3 and not the best possible ones (see [2]).

Other quadratic spline approaches based on function values given at the mesh points have been studied by Kammerer et al. [3], M.J. Marsden [4], S. Demko [1], E. Neuman [5] and Sharma and Tzimbalario [6], among many others.

2. THE INTERPOLATION PROBLEM.

Let $\{x_i, i=0,1,\ldots,N+1\}$ be a uniform partition of $[0,1]$ with $h=x_{i+1}-x_i$ $(i=0,1,\ldots,N)$. Denote by $S^{(1)}_{N,2}$ the linear space of quadratic splines $s(x)$ such that:

$$\begin{cases} s(x) \in C^1[0,1], \\ s(x) \text{ is a quadratic polynomial in each subinterval } [x_i,x_{i+1}]. \end{cases}$$

If $g:[0,1] \to \mathbb{R}$ is a real -valued function and $\lambda \in \mathbb{R}(0\leq\lambda\leq 1)$ is a real parameter then $g^{(k)}_{i+\lambda}$ stands for $g^{(k)}(x_i + \lambda h)(i=0,1,\ldots,N)$; $k \geq 0$.

We now state and prove our first result.

THEOREM 1:

Given the arbitrary real numbers $f'_{i+\lambda}(i=0,1,\ldots N)$, f_0 and f_{N+1}, then there exists a unique quadratic spline $s \in S^{(1)}_{N,2}$ such that

$$s'_{i+\lambda} = f'_{i+\lambda}, \quad (i=0,1,\ldots,N)$$

and (2.1)

$$s_0 = f_0, \quad s_{N+1} = f_{N+1}$$

whenever $0 \leq \lambda \leq 1$ and $\lambda \neq 1/2$.

---

1) Kuwait University, Department of Mathematics, P.O. Box 5969 - 13060 Kuwait

PROOF.

If $P_\lambda(t)$ is a quadratic polynomial in $[0,1]$, then it can be expressed as

$$p_\lambda(t) = p_\lambda(0)\, A_\lambda(t) + p'_\lambda(\lambda)\, B_\lambda(t) + p_\lambda(1)\, C_\lambda(t)$$

where

$$\begin{aligned} A_\lambda(t) &= (t^2 - 2\lambda t)/(2\lambda-1) + 1, \\ B_\lambda(t) &= (t^2 - t)/(2\lambda-1), \\ C_\lambda(t) &= -(t^2 - 2\lambda t)/(2\lambda-1), \end{aligned} \qquad (2.2)$$

provided $\lambda \neq 1/2$.

Fix $i \in \{1,2,\dots,N\}$, set $x = x_i + th$, $0 \leq t \leq 1$, then the quadratic spline $s(x)$ which satisfies (2.1) in $[x_i, x_{i+1}]$ will be

$$s(x) = s_i A_\lambda(t) + hf'_{i+\lambda} B_\lambda(t) + s_{i+1} C_\lambda(t) \qquad (2.3)$$

with a similar expression for $s(x)$ in $[x_{i-1}, x_i]$. Since $s \in C^1[0,1]$ then $s'(x_i^-) = s'(x_i^+)$ $(i = 1,2,\dots,N)$ will lead to the following linear system of equations:

$$\alpha s_{i-1} + s_i + \beta s_{i+1} = \frac{1}{2(1-2\lambda)}\, h\, (f'_{i-1+\lambda} + f'_{i+\lambda}) \qquad (2.4)$$

$(i=1,2,\dots,N)$, where

$$\alpha = \frac{\lambda-1}{2\lambda-1} \quad \text{and} \quad \beta = \frac{-\lambda}{2\lambda-1} .$$

The coefficient matrix M_λ of this system has the form

$$\begin{bmatrix} 1 & \beta & 0 & \cdots & 0 \\ \alpha & 1 & \beta & \ddots & 0 \\ 0 & \ddots & \ddots & \ddots & \beta \\ 0 & \cdots & 0 & \alpha & 1 \end{bmatrix}$$

Denote by δ_N the determinant of this matrix. It is not difficult to show that the following recurrence relation holds

$$\delta_N = \delta_{N-1} - \alpha\beta\delta_{N-2} ,$$
$$\delta_1 = 1 \quad \text{and} \quad \delta_2 = 1 - \alpha\beta$$

but $\alpha\beta \leq 0$ for all $0 \leq \lambda \leq 1$. It follows then (since $\delta_1 > 0$ and $\delta_2 > 0$) that $\delta_N > 0$. Hence the matrix M_λ is nonsingular and $s(x)$ is uniquely determined.

3. ERROR IN INTERPOLATION.

In this section an L_∞ error estimate for the above interpolatory quadratic spline will be presented. It is useful to introduce the following indentities [see (2.2)]:

$$\begin{aligned} &A_\lambda(t) + C_\lambda(t) = 1 , \quad B_\lambda(t) + C_\lambda(t) = t \\ &2\lambda B'_\lambda(t) + C'_\lambda(t) = 2t , \quad (2\lambda-1) B'_\lambda(t) = (2t-1) . \end{aligned} \qquad (3.1)$$

The main result of this section is the following theorem.

THEOREM 2:

Let $s(x)$ be the quadratic spline defined in 2. If $f \in C^\ell[0,1]$, $\ell = 2$ or 3, then

$$\| s'-f' \|_\infty \leq \begin{cases} \left[\dfrac{\theta}{(1-2\lambda)^2}(1+h) + \dfrac{\theta h}{4|1-2\lambda|} + \dfrac{1}{2} \right] \omega_2(h) , & \ell = 2 \\[2ex] \left[\dfrac{\theta}{6(1-2\lambda)^2} (h+h^2)(6\lambda^2-6\lambda+5) \right. \\ \left. \quad + \dfrac{\theta}{12} \left(\dfrac{3|3\lambda-1|+1}{|2\lambda-1|} \right) h^2 \right] \| f^{(3)} \|_\infty , & \ell = 3, \end{cases} \qquad (3.2.1)$$

and

$$\|s - f\|_\infty \le \frac{1}{2} \|s' - f'\|_\infty \tag{3.2.2}$$

where $\omega_2(h)$ denotes the modulus of continuity of $f^{(2)}$, $[\omega_2(h) \le 2\|f^{(2)}\|_\infty]$ and

$$\theta = \begin{cases} 1-\lambda \ , & \text{if } \ 0 \leqq \lambda < 1/2 \\ \lambda \ , & \text{if } \ 1/2 < \lambda \leqq 1 \end{cases}$$

To prove this theorem we need the following lemma.

<u>LEMMA 1:</u>

If $f \in C^{\ell}[0,1]$, $\ell = 2$ or 3 then we have $(e_i = s(x_i)-f(x_i),\ i = 0,1,2,\dots,N)$

$$|e_{i+1} - e_i| \le \begin{cases} \dfrac{1}{2|1-2\lambda|}(\gamma h + h^2)\ \omega_2(h)\ , & \ell = 2 \\[2ex] \dfrac{1}{12|1-2\lambda|}(\gamma h^2 + h^3)(6\lambda^2 - 6\lambda + 5)\ \|f^{(3)}\|_\infty\ , & \ell = 3 \end{cases} \tag{3.3}$$

where $\gamma = (1-\theta)/\theta$.

<u>PROOF.</u>

For the sake of brevity, we will consider only the case $\ell = 2$. From (2.4) we have $(i = 1,2,\dots,N\ ,\ e_0 = e_{N+1} = 0)$

$$-(1-\lambda)e_{i-1} + (1-2\lambda)e_i + \lambda e_{i+1} = \frac{h}{2}(f'_{i-1+\lambda} + f'_{i+\lambda}) + (1-\lambda)\ f_{i-1} - (1-2\lambda)f_i - \lambda f_{i+1} \tag{3.4}$$

Now let

$$z_i = e_{i+1} - e_i\ ,\ i = 0,1,\dots,N \tag{3.5}$$

and expanding the right-hand side of (3.4) about x_i using Taylor's expansion of order ℓ, we get

$$(1-\lambda)z_{i-1} + \lambda z_i = b_i\ (i = 1,2,\dots,N) \tag{3.6}$$

where

$$b_i = \frac{h^2}{2}\Big\{(1-\lambda)[f''(\xi_{i,1} - f''(\xi_{i,2})] + \lambda[f''(\xi_{i,3}) - f''(\xi_{i,4})]\Big\}, \tag{3.7}$$

and $\xi_{i,1}$, $\xi_{i,2} \in (x_{i-1},x_i)$ and $\xi_{i,3}, \xi_{i,4} \in (x_i,x_{i+1})$.

First we consider the case $\frac{1}{2} < \lambda \le 1$. In this case $\gamma = (1-\lambda)/\lambda$, thus $0 \le \gamma < 1$ and the system (3.6) can be written in the lower bidiagonal form

$$\begin{bmatrix} 1 & & 0 & \cdots & 0 \\ \gamma & 1 & 0 & \cdots & 0 \\ 0 & \ddots & \ddots & \ddots & \vdots \\ \vdots & \ddots & \ddots & \ddots & 0 \\ 0 & \cdots & 0 & \gamma & 1 \end{bmatrix} \begin{bmatrix} z_1 \\ z_2 \\ \cdot \\ \cdot \\ \cdot \\ z_N \end{bmatrix} = \begin{bmatrix} c_1-\gamma z_0 \\ c_2 \\ \cdot \\ \cdot \\ \cdot \\ c_N \end{bmatrix}$$

where

$$c_i = b_i/\lambda \ - \ (i = 1,2,\dots,N). \tag{3.8}$$

Therefore

$$\begin{bmatrix} z_1 \\ z_2 \\ \cdot \\ \cdot \\ \cdot \\ z_N \end{bmatrix} = \begin{bmatrix} 1 & 0 & \cdots & & 0 \\ \beta & 1 & 0 & \cdots & 0 \\ \beta^2 & \ddots & \ddots & \ddots & 0 \\ \vdots & & \ddots & \ddots & \\ \beta^{N-1} & \cdots & \beta^2 & \beta & 1 \end{bmatrix} \begin{bmatrix} c_i + z_0 \\ c_2 \\ \cdot \\ \cdot \\ \cdot \\ c_N \end{bmatrix} , \beta = -\gamma \qquad (3.9)$$

Hence

$$z_i = [\beta^i z_0 + \sum_{j=1}^{i} \beta^{i-j} c_j] \quad , \quad (i = 1,2,\ldots,N) \qquad (3.10)$$

Since

$$\sum_{i=0}^{N} z_i = \sum_{i=0}^{N} (e_{i+1} - e_i) = e_{N+1} - e_0 = 0$$

Then

$$\sum_{i=1}^{N} z_i = - z_0 \; . \qquad (3.11)$$

From (3.9) and (3.11) we obtain

$$- z_0 = z_0 \sum_{i=1}^{N} \beta^i + c_1 \sum_{i=0}^{N-1} \beta^i + c_2 \sum_{i=0}^{N-2} \beta^i + \ldots + c_{N-1} \sum_{i=0}^{1} \beta^i + c_N$$

Thus

$$(1-\beta^{N+1})(-z_0) = c_1(1-\beta^N) + c_2(1-\beta^{N-1}) + \ldots + c_N(1-\beta) \; .$$

Since $(1-\beta^{N+1}) \geq 1-\beta^2 > 0$ and $1-\beta^i \leq 1-\beta$, $(i=1,2,\ldots,N)$, then

$$|z_0| \leq Nc\ (1-\beta)/(1-\beta^2) = Nc\lambda/(2\lambda-1) = Nb/2\lambda-1)$$

where $c = \max_{1\leq i\leq N} |c_i| = b/\lambda$, $b = \max_{1\leq i\leq N} |b_i|$.

Thus, using (3.10), we have

$$|z_0| \leq Nb/2\lambda-1) \leq \frac{b}{h(2\lambda-1)}$$

and (3.12)

$$|z_i| \leq \frac{1-\lambda}{\lambda} |z_0| + \frac{b}{(2\lambda-1)} \qquad (i=1,2,\ldots,N) \; .$$

Therefore

$$|z_i| \leq \frac{b}{h(2\lambda-1)} + \frac{b}{2\lambda-1} \quad (i = 0,1,\ldots,N) \qquad (3.13)$$

It follows then, from (3.7) and (3.12), that

$$|z_i| \leq \frac{1-\lambda}{\lambda} \cdot \frac{h\omega_2(h)}{2(2\lambda-1)} + \frac{h^2\omega_2(h)}{2(2\lambda-1)} \; , \quad (i = 0,1,2,\ldots N) \qquad (3.14)$$

which proves the first inequality of (3.3).

Now we consider the case $0 \leq \lambda \leq \frac{1}{2}$. We find in a similar manner, where the system (3.6) will have an upper-bidiagonal form, that

$$|z_{N+1}| \leq Nb/(1-2\lambda) \; ,$$

and **(3.15)**

$$|z_i| \leq \frac{\lambda}{1-\lambda} |z_{N+1}| + \frac{b}{1-2\lambda} \; , \quad (i = 1,2,\ldots,N).$$

This completes the proof of Lemma 1.

Now we are in a position to prove theorem 2.

PROOF OF THEOREM 2:

Again we consider the case where $f \in C^2[0,1]$. The other case is similar. Letting $x \in [x_i, x_{i+1}]$, $(i = 0,1,2,\ldots,N)$ using (2.3), we get

$$h[s'(x) - f'(x)] = e_i A'_\lambda(t) + hf'_{i+\lambda} B'_\lambda(t) + e_{i+1} C'_\lambda(t) + f_i A'_\lambda(t) + f_{i+1} C'_\lambda(t) - hf'_\lambda(x)$$

which, using (3.1), gives

$$h[s'(x) - f'(x)] = (e_i - e_{i+1})A'_\lambda(t) + (f_i - f_{i+1})A'_\lambda(t) + hf'_{i+\lambda}B'_\lambda(t) - hf'_\lambda(x)$$

Expanding the right hand side using Taylor's expansion of order two about $x_i + h/2$, we get

$$h[s'(x) - f'(x)] = (e_i - e_{i+1})A'_\lambda(t) + \left[\frac{h^2}{8} f''(\alpha_i) - \frac{h^2}{8} f''(\beta_i)\right] A'_\lambda(t) + (\lambda - \tfrac{1}{2})h^2\left[f''(\gamma_i) - f''(\delta_i)\right] B'_\lambda(t) \tag{3.16}$$

where $\alpha_i, \beta_i, \gamma_i$ and $\delta_i \in (x_i, x_{i+1})$.

But from (2.2) we have

$$|A'_\lambda(t)| \le \frac{2\theta}{|1-2\lambda|},$$

and

$$|B'_\lambda(t)| \le \frac{1}{|1-2\lambda|}$$

Using this, together with (3.16), we get

$$h|s'(x) - f'(x)| \le |e_i - e_{i+1}| \frac{2\theta}{|1-2\lambda|} + \frac{h^2}{4}\omega_2(h)\frac{\theta}{|1-2\lambda|} + \frac{h^2}{2}\omega_2(h) \tag{3.17}$$

This inequality with (3.3) gives

$$|s'(x) - f'(x)| \le \left[\frac{\theta}{(1-2\lambda)^2}(1+h) + \frac{\theta}{4|1-2\lambda|}h + \frac{h}{2}\right]\omega_2(h) \tag{3.18}$$

This proves (3.2.1). We integrate $s'(t) - f'(t)$ over $[0,x]$ (resp. over $[x,1]$) if x is closer to 0 (resp. to 1), using $s(0) - f(0) = 0$ (resp. $s(1) - f(1) = 0$) and (3.18) and then (3.2.2) follows. This completes the proof of Theorem 2.

4. THE PARTICULAR CASES $\lambda = 0$ and $\lambda = 1$

Suppose $\lambda = 0$ (the case $\lambda = 1$ is very similar) then (2.4) is reduced to the recurrence formula $(i = 1,2,\ldots,N)$

$$s_i = s_{i-1} + \frac{h}{2}(f'_{i-1} + f'_i) \tag{4.1}$$

and (2.3) takes the form

$$s(x) = (-t^2+1)s_i + (-t^2 + t)hf'_i + t^2 s_{i+1} \tag{4.2}$$

with $t = (x-x_i)/h$. From (4.1) we have $(i = 1,2,\ldots,N)$

$$s_i = s_0 + \frac{h}{2}[f'_0 + f'_i + 2\sum_{j=1}^{i-1} f'_j]$$

Therefore

$$s_i - f_i = \frac{h}{2}[f'_0 + f'_i + 2\sum_{j=1}^{i-1} f'_j] - \int_0^{x_i} f'(x)\,dx$$

since $s_0 = f_0$ The right hand side is in fact the error of the classical trapesoidal rule applied to the function $f'(x)$ in $[0,x_i]$, Thus

$$|s_i - f_i| \le \frac{h^2}{12} \|f^{(3)}\|_\infty \quad , \quad (i = 0,1,\dots,N+1) \tag{4.3}$$

provided $f \in C^3[0,1]$. Moreover, in $[x_i,x_{i+1}]$, $i(i = 0,1,\dots,N-1)$, $s'(x)$ is the linear Lagrange interpolant for f', hence

$$|s'(x) - f'(x)| \le \frac{h^2}{8} \|f^{(3)}\|_\infty \ ,\forall\ x \text{ in } [0,x_N] \tag{4.4}$$

provided $f \in C^3[0,1]$. Integrating over $[0,x_N]$ using $s(0) - f(0) = 0$, we obtain

$$|s(x) - f(x)| \le \frac{h^2}{8} \|f^{(3)}\|_\infty \ ,\forall\ x \text{ in } [0,x_N] \tag{4.5}$$

In the last subinterval $[x_N,x_{N+1}]$, differentiating (4.2) with respect to x twice and subtracting $f''(x)$ from both sides, we get

$$h^2(s''(x) - f''(x)) = -2(s_N-f_N) - 2hf'_N + 2f_{N+1} - 2f_N - h^2f''(x)$$

Expanding the right hand side about $x_N + h/2$, we have

$$h^2|s''(x) - f''(x)| \le 2|s_N - f_N| + \frac{5}{6} h^3 \|f^{(3)}\|_\infty .$$

Thus using (4.3), we have

$$|s''(x) - f''(x)| \le \frac{1}{6} (1 + 5h) \|f^{(3)}\|_\infty$$

Also, differentiating (4.2) and subtracting $f'(x)$ from both sides, we get

$$h[s'(x) - f'(x)] = 2t(s_N - f_N) - 2tf_N + (-2t + 1) hf'_N + 2tf_{N+1} - hf'(x)$$

Expanding the right hand side about $x_N + h/2$, we obtain

$$h|s'(x) - f'(x)| \le 2|s_N - f_N| + \frac{1}{3} h^3 \|f^{(3)}\|_\infty$$

Therefore, using (4.3),

$$|s'(x) - f'(x)| \le \frac{1}{6} h (1 + 2h) \|f^{(3)}\|_\infty$$

Integrating over $[x_N,x_{N+1}]$ and using $s_{N+1} = f_{N+1}$, we get

$$|s(x) - f(x)| \le \frac{1}{6} h^2 (1 + 2h) \|f^{(3)}\|_\infty$$

REFERENCES

[1] S. DEMKO, Interpolation by quadratic splines, J. Approx. Theory 23 (1978),392-400.

[2] M.N. El-TARAZI AND S. SALLAM, On quartic splines with application to quadratures, Computing 38 (1987), 355-361.

[3] W.J. KAMMERER, G.W. REDDIEN, AND R.S. VARGA, Quadratic interpolatory splines, Numer. Math. 22 (1979), 241-259.

[4] M.J. MARSDEN, Quadratic spline interpolation, Bull. Amer. Math. Soc. 80 (1974), 903-906.

[5] E. NEUMAN, Quadratic splines and histospline projection, J. Approx. Theorey 29 (1980), 297-304.

[6] A. SHARMA AND J. TZIMBALARIO, Quadratic splines, J. Approx. Theory 19 (1977), 186-193

COMPLEX SPLINES AND SINGULAR INTEGRAL EQUATIONS

Gunther Schmidt [1)]

Complex analysis has various applications in applied sciences. But for many problems one cannot get the explicit form of the solutions and has to construct approximating functions from the given conditions. One possibility of the construction of such functions is the use of splines in the complex plane. To be more precisely, on a given curve one considers piecewise polynomials of the complex variable, the so-called complex splines, and Cauchy typ integrals of complex splines, the analytic splines. There are several papers dealing with interpolation properties of complex and analytic splines (cf. [1,2,4,12,14,15]), with the approximation of complex functions on curves, analytic and harmonic functions in domains (cf. [3,7,8,16]). The numerical solution of twodimensional boundary value problems using complex and analytic splines is described in [9].

The present paper deals with the application of complex splines to the approximate solution of singular integro-differential equations on closed curves, which have several application in mechanics (cf. [10]). We show that known results about the convergence of approximation methods with real polynomial splines for pseudodifferential equations on closed curves are also valid for the case when complex splines are used as trial functions. Moreover, we give recurrence relations for Cauchy typ integrals and singular integrals with Cauchy kernel of complex B-splines and for their derivatives, which allow an exact and effective evaluation of point values of these integrals.

1. Let Γ be a simple closed C^∞-curve given by a parametrization $z=\gamma(x)$, $x\in[0,1]$. We identify functions on Γ with 1-periodic functions by $K_\gamma u(x)=u(\gamma(x))$, $x\in\mathbb{R}$. H^s, $s\in\mathbb{R}$, denotes the periodic Sobolev space on $\mathbb{R}$, that means the closure of the space of smooth 1-periodic functions C^∞_{per} with respect to the norm

$$\|u\|_s := (|\hat{u}_0|^2 + \sum_{0\neq k\in\mathbb{Z}} |\hat{u}_k|^2|2\pi k|^{2s})^{1/2} \quad , \quad \hat{u}_k = \int_0^1 u(x)e^{-2\pi ik}dx.$$

By $\Delta = \{x_k\}_{k=-\infty}^{\infty}$ we denote a 1-periodic partition of $\mathbb{R}$ with $x_j < x_{j+1}$, $x_{j+n} = x_j+1$, $\forall\, j\in\mathbb{Z}$, and $0\le x_0 < x_{n-1} < 1$. We introduce for integer $d\ge 0$ the linear spaces

$$S^d_\Delta = \left\{u\in C^{d-1}_{per}:\ u(x) = \sum_{k=0}^{d} a_{jk}x^k \ , x_j < x < x_{j+1}\right\} \quad ,$$

---

1) Karl-Weierstraß-Institut für Mathematik der AdW der DDR
DDR - 1086 Berlin, Mohrenstr.39

$$\Gamma_\Delta^{d} = \left\{ u \in C_{per}^{d-1} : u(x) = \sum_{k=0}^{d} a_{jk}(\gamma(x))^k \ , \ x_j < x < x_{j+1} \right\} ,$$

($C_{per}^{-1} = L_{per}^{\infty}$). S_Δ^d is the well-known set of 1 - periodic polynomial splines of degree d and defect 1. Obviously, $P_{\Delta'}^d := K_\gamma^{-1}(\Gamma_\Delta^d)$ is the linear space of complex splines on Γ subordinate to the partition $\Delta' = \{ z_k = \gamma(x_k) \}_{k=0}^{n-1}$.

For the following we need the linear operator D, defined by

$$Du(x) := \frac{1}{i}\frac{du}{dx} + \int_0^1 u(t)dt = \frac{1}{i}\frac{du}{dx} + Ju \ , \tag{1.1}$$

which can be extended to an invertible operator $D : H^s \longrightarrow H^{s-1}$, $s \in \mathbb{R}$. Furthermore we define

$$\Gamma_\Delta^{0,d} = \left\{ u_\Delta \in S_\Delta^0 : \int_0^1 u_\Delta(x)\gamma'(x)(\gamma(x))^k dx = 0 \ , \ k = 0, \dots, d-1 \right\}$$

and

$$B_d = \operatorname{span} \left\{ \left(\frac{1}{\gamma'}\frac{d}{dx}\right)^k \frac{1}{\gamma'(x)} \ , \ k = 0, \dots, d-1 \right\} .$$

<u>Lemma 1:</u> For any $d \in \mathbb{N}$ there holds $S_\Delta^d = D^{-d}(S_\Delta^0)$.

<u>Proof:</u> See [5].

<u>Lemma 2:</u> For any $d \in \mathbb{N}$ there holds $\Gamma_\Delta^d = (D^{-1}\gamma')^d(\Gamma_\Delta^{0,d} \dotplus B_d)$.

<u>Proof:</u> By (1.1)

$$\left(\frac{1}{\gamma'} D\right)^d = \left(\frac{1}{i\gamma'}\frac{d}{dx}\right)^d + B \ ,$$

where B is the sum of products of d operators $\frac{1}{\gamma'}J$ or $\frac{1}{i\gamma'}\frac{d}{dx}$. Since the image of J is the set of constant functions we have im B = B_d. Furthermore, $\left(\frac{1}{\gamma'}\frac{d}{dx}\right)^d$ maps polynomials of degree d in $\gamma(x)$ into constant functions, therefore $\left(\frac{1}{i\gamma'}\frac{d}{dx}\right)^d u_\Delta \in S_\Delta^0$ for $u_\Delta \in \Gamma_\Delta^d$. The operator $\left(\frac{1}{i\gamma'}\frac{d}{dx}\right)^d : H^{d/2} \longrightarrow H^{-d/2}$ is bounded, its adjoint is $\left(\frac{1}{i}\frac{d}{dx}\frac{1}{\overline{\gamma'}}\right)^d$ with kernel

$$\ker \left(\frac{1}{i}\frac{1}{dx}\frac{1}{\overline{\gamma'}}\right)^d = \operatorname{span}\left\{ \overline{\gamma'\gamma^k} \ , \ k = 0, \dots, d-1 \right\} .$$

Therefore

$$\left(\frac{1}{i\gamma'}\frac{d}{dx}\right)^d (\Gamma_\Delta^d) = \left\{ u_\Delta \in S_\Delta^0 : u_\Delta \perp \operatorname{span} \overline{\gamma'\gamma^k}, \ k = 0, \dots, d-1 \right\} = \Gamma_\Delta^{0,d} . \blacksquare$$

Using Lemma 2 and known approximation estimates for piecewise constant functions we obtain the same approximation properties as for S_Δ^d.

<u>Lemma 3:</u> For any $s < d + 1/2$ and $s \le r \le d + 1$, $u \in H^s$ there exists $u_\Delta \in \Gamma_\Delta^d$ such that

$$\| u - u_\Delta \|_t \le c \bar{\Delta}^{\,r-t} \| u \|_r$$

for all $t \le s$, c does not depend on u and Δ ($\bar{\Delta} = \max_j (x_j - x_{j-1})$).

<u>Proof:</u> See [13].

Lemma 4: Let $r < 1/2$. For any Δ and $v_\Delta \in \Gamma_\Delta^{0,d} \dot{+} B_d$ there exists $u_\Delta \in S_\Delta^0$ such that for all $t \le r$ and $s < 1/2$

$$\| u_\Delta - v_\Delta \|_t \le c \bar{\Delta}^{1-t} \| v_\Delta \|_s ,$$

c does not depend on Δ and v_Δ.

Proof: We note that

$$((\frac{1}{\gamma'}\frac{d}{dx})^k \frac{1}{\gamma'}, \overline{\gamma'\gamma^j}) = \begin{cases} (-1)^k \frac{j!}{(j-k)!} \int_0^1 \gamma^{j-k}\, dx , & j \ge k \\ 0 , & j < k , \end{cases}$$

$(\cdot , \cdot)$ denotes the usual L^2 scalar product. Therefore there exists a projection operator P such that $\operatorname{im} P = B_d$ and $((I-P)u, \overline{\gamma'\gamma^j}) = 0$, $j = 0,\dots,d-1$. Let $v_\Delta \in \Gamma_\Delta^{0,d} \dot{+} B_d$. There exists $w_\Delta \in S_\Delta^0$ such that

$$\| Pv_\Delta - w_\Delta \|_t \le c \bar{\Delta}^{1-t} \| Pv_\Delta \|_1 \quad \text{for all } t \le r.$$

Now $w_\Delta + (I-P)v_\Delta \in S_\Delta^0$ and

$$\| v_\Delta - (w_\Delta + (I-P)v_\Delta \|_t = \| Pv_\Delta - w_\Delta \|_t \le c \bar{\Delta}^{1-t} \| Pv_\Delta \|_1$$
$$\le c \bar{\Delta}^{1-t} (\sup_{\|Pu\|_s \le 1} \| Pu \|_1) \| Pv_\Delta \|_s .$$

Since $\dim B_d = d$ we have $\sup_{\|Pu\|_s \le 1} \|Pu\|_1 \le c_1$. ∎

Analog ously we can prove

Lemma 5: For any $u_\Delta \in S_\Delta^0$ and $\rho < r < 1/2$ there exists $v_\Delta \in \Gamma_\Delta^{0,d} \dot{+} B_d$ such that for all $t \le r$, $\rho \le s < 1/2$

$$\| u_\Delta - v_\Delta \|_t \le c \bar{\Delta}^{1-t} \| u_\Delta \|_s .$$

2. Let us given the equation

$$A u = f \tag{2.1}$$

where A is a pseudodifferential operator defined for $u \in C_{per}^\infty$ by

$$Au(x) = \sum_{0 \ne k \in \mathbb{Z}} \sigma(x, 2\pi k) \hat{u}_k e^{2\pi i k x} + \int_0^1 k(x,t) u(t)\, dt.$$

Here the function $\sigma(x,\xi)$ is 1-periodic in x, smooth for all x and $\xi \ne 0$. Furthermore it admits an asymptotic representation

$$\sigma \sim \sigma_m + \sigma_{m-1} + \dots$$

where the functions σ_j, $j \le m$, are positive homogeneous of degree j in ξ, $k(x,t)$ is a 1-periodic smooth kernel function. A extends to a bounded operator $H^s \to H^{s-m}$, $s \in \mathbb{R}$, the function σ_m is called the principal symbol and m the order of A (for details see [6], e.g.). We note that D is a pseudodifferential operator with the symbol ξ.

In the last years various approximation methods for solving pseudodifferential equations (2.1) are studied, where the solution is approximated by splines of the class S_Δ^d (cf. [11,13]). There necessary and sufficient conditions for the stability of these methods were obtained, which ensure the convergence of the approximate solutions to the exact solution with optimal order if $\bar{\Delta} \to 0$. Using Lemmas 4 and 5 it can be proved that the same conditions ensure the convergence of the corresponding approximation methods if the approximate solution is sought in the class Γ_Δ^d.

As the simplest example we consider the nodal collocation:

Find $u_\Delta \in X_\Delta$ ($X_\Delta = S_\Delta^d$ or $X = \Gamma_\Delta^d$) such that

$$Au_\Delta(x_k) = f(x_k) \quad , \; k = 0,\dots,n-1. \tag{2.2}$$

The approximate equations make sense iff $d > m$. We list some results obtained for the case $X_\Delta = S_\Delta^d$ (cf. [5,11]).

1. There exist interpolation projections Q_Δ with the properties $Q_\Delta f(x_k) = f(x_k)$, $k = 0,\dots,n-1$, $\dim \operatorname{im} Q_\Delta = n$ and for all $f \in H^s$

$$\| Q_\Delta f - f \|_t \le c \bar{\Delta}^{s-t} \| f \|_s ,$$

where $0 \le t \le (d-m+1)/2 \le s \le d-m+1$ for arbitrary Δ and $0 \le t \le s \le d-m+1$, $s > 1/2$, $t < d-m+1/2$ for quasiuniform partitions Δ, i.e. there exists a constant ϱ such that $\min(x_j - x_{j-1}) \ge \varrho \bar{\Delta}$ for all Δ of a given class.

Using these projections equations (2.2) can be written as

$$Q_\Delta A u_\Delta = Q_\Delta f \quad . \tag{2.3}$$

2. The operators $Q_\Delta A|_{S_\Delta^d}$ are stable in (H^s, H^{s-m}), i.e., there exist positive constants c, h_0 such that for all $u_\Delta \in S_\Delta^d$ with $\bar{\Delta} \le h_0$

$$\| Q_\Delta A u_\Delta \|_{s-m} \ge c \| u_\Delta \|_s ,$$

if and only if either d is odd and the quotient $\sigma_m(x,+1)/\sigma_m(x,-1) \notin (-\infty, 0]$ or d is even and $\sigma_m(x,+1)/\sigma_m(x,-1) \notin [0, \infty)$. This holds for $s = (d+m-1)/2$ for arbitrary partitions Δ and $m+1/2 < s < d+1/2$ for quasiuniform Δ.

Theorem 1: The operators $Q_\Delta A|_{\Gamma_\Delta^d}$ are stable if and only if $Q_\Delta A|_{S_\Delta^d}$ are stable.

Proof: Assume that $\| Q_\Delta A u_\Delta \|_{s-m} \ge c \| u_\Delta \|_s$ for all $u_\Delta \in S_\Delta^d$, $\bar{\Delta} \le h_0$. For $v_\Delta \in \Gamma_\Delta^d$ there exists $\tilde{v}_\Delta \in \Gamma_\Delta^{0,d} \dotplus B_d$ such that $v_\Delta = (D^{-1}\gamma')^d \tilde{v}_\Delta$. By Lemma 4 we choose $\tilde{u}_\Delta \in S_\Delta^0$ with $\| \tilde{v}_\Delta - \tilde{u}_\Delta \|_{s-d} \le c \bar{\Delta}^{1-s+d} \| \tilde{v}_\Delta \|_{s-d}$. Then

$$Q_\Delta A v_\Delta = Q_\Delta A \, (D^{-1}\gamma')^d \tilde{u}_\Delta + Q_\Delta A (D^{-1}\gamma')^d (\tilde{v}_\Delta - \tilde{u}_\Delta)$$

and

$$\| Q_\Delta A v_\Delta \|_{s-m} \ge \| Q_\Delta A (D^{-1}\gamma')^d \tilde{u}_\Delta \|_{s-m} - c \bar{\Delta}^{1-s+d} \| \tilde{v}_\Delta \|_{s-d} .$$

Now $\tilde{u}_\Delta = D^d u_\Delta$ for some $u_\Delta \in S_\Delta^d$ and $Q_\Delta A (D^{-1}\gamma')^d \tilde{u}_\Delta = Q A (D^{-1}\gamma')^d D^d u_\Delta$. The operator $T := A(D^{-1}\gamma')^d D^d - (\gamma')^d A$ is a pseudodifferential operator of order $m-1$ (cf.[6]), hence $T: H^s \to H^{s-m}$ is compact. Thus we have

$$Q_\Delta A (D^{-1}\gamma')^d D^d u_\Delta = Q_\Delta (\gamma')^d A u_\Delta + Q_\Delta T u_\Delta .$$

Since Q_Δ is an interpolation projection we have

$$Q_\Delta (\gamma')^d A u_\Delta = Q_\Delta (\gamma')^d Q_\Delta A u_\Delta .$$

Because of $\gamma' \neq 0$ the stability of $Q_\Delta A\big|_{S_\Delta^d}$ implies the stability of $Q_\Delta(\gamma') A\big|_{S_\Delta^d}$. Furthermore, $(\gamma')^d A + T = A(D^{-1}\gamma')^d D^d$ is invertible, standard results on projection methods yield the stability of $Q_\Delta A (D^{-1}\gamma')^d D^d\big|_{S_\Delta^d}$. Therefore

$$\|Q_\Delta A v_\Delta\|_{s-m} \geq c\|u_\Delta\|_s - c\bar{\Delta}^{1-s+d}\|v_\Delta\|_s .$$

The inequality

$$\|u_\Delta\|_s = \|\tilde{u}_\Delta\|_{s-d} \geq c_1(1-\bar{\Delta}^{1-s+d})\|\tilde{v}_\Delta\|_{s-d} \geq c_2(1-\bar{\Delta}^{1-s+d})\|v_\Delta\|_s$$

proves the stability of $Q_\Delta A\big|_{\Gamma_\Delta^d}$. Using Lemma 5 the other direction can be proved quiet analogou sly. ∎

Using the stability and the properties of the projections Q_Δ it is easy to prove

<u>Theorem 2:</u> Let A be an invertible pseudodifferential operator of order m, $d > m$ a nonnegative integer. Assume that the principal symbol of A satiesfies $\sigma_m(x,+1)/\sigma_m(x,-1) \notin (-\infty,0]$ if d is odd, resp. $\sigma_m(x,+1)/\sigma_m(x,-1) \notin [0,\infty)$ if d is even. Then there exists h_0 such that for all partitions Δ with $\bar{\Delta} \leq h_0$ and $u_\Delta \in \Gamma_\Delta^d$ the approximate equations (2.2) are uniquely solvable. If $f \in H^{s-m}$, $(d+1+m)/2 \leq s \leq d+1$, then the approximate solutions $u_\Delta \in \Gamma_\Delta^d$ converge to the exact solution u of (2.1) with

$$\|u - u_\Delta\|_t \leq c\bar{\Delta}^{s-t}\|f\|_{s-m} \qquad (2.4)$$

for $m \leq t \leq (d+1+m)/2$. If the considered partitions arequasiuniform then (2.4) holds for $m \leq t \leq s \leq d+1$, $s > m+1/2$, $t < d+1/2$.

The same methods as in the proof of Theorem 1 allow to give conditions for the convergence of other approximation methods, as Galerkin methods or collocation with collocation points different from the mesh points (cf.[13]).

<u>3.</u> As a special case of (2.1) we consider a singular integro- differential equation on Γ

$$\mathcal{A}\, u(z) = \sum_{l=0}^{m} \left(a_l(z)u^{(l)}(z) + \frac{b_l(z)}{\pi i}\int_\Gamma \frac{u^{(l)}(\zeta)}{\zeta - z}\, d\zeta\right) = f(z) , \qquad (3.1)$$

$z \in \Gamma$. For a given parametrization $z = \gamma(x)$ of Γ $A = K_\gamma \mathcal{A} K_\gamma^{-1}$ is a pseudodifferential operator of order m with principal symbol

$$\sigma_m(x,\xi) = (a_m(\gamma(x)) + b_m(\gamma(x))\,\xi/|\xi|\,)\,(i\xi)^m\,(\gamma')^{-m}\ .$$

The approximate solution of (3.1) with spline methods requires the evaluation of singular integrals of piecewise polynomial functions. If these functions are piecewise polynomials of the arc length the integrals must be evaluated using quadrature methods. If the trial functions are complex splines these integrals can be evaluated exactly.

Let us consider complex B-splines on the contour Γ. There is given a partition Δ' of n points $z_0,\dots,z_{n-1}$ arranged in counterclockwise order separating Γ into closed arcs $\Gamma_j, j=0,\dots,n-1$. For a given arc $\Gamma'\subset\Gamma$ and points $z,\zeta\in\Gamma'$ we denote $z\prec\zeta$ if a point running in counterclockwise order on Γ' meets point z, first, then point ζ. For nonnegative integer $d\leq n-2$ we define the function

$$(\zeta - z)_+^d = \begin{cases} (\zeta - z)^d & \text{if } z,\zeta\in \bigcup_{k=j}^{j+d}\Gamma_k \text{ for some } j \text{ and } z\prec\zeta, \\ 0 & \text{for all other cases}. \end{cases}$$

Here and in the following $\Gamma_{n+j}=\Gamma_j$, $z_{n+j}=z_j$, $j=0,\dots,n-1$. Then the complex B-splines are defined as the divided differences

$$M_{j,d}(z) = (d+1)\cdot[z_j,\dots,z_{j+d+1};(\cdot - z)_+^d]\quad,\ j=0,\dots,n-1,$$

(cf.[7]). The following properties are well known:

- $\operatorname{supp} M_{j,d} = \bigcup_{k=j}^{k+j}\Gamma_k$
- $\operatorname{span}\{M_{j,d},\ j=0,\dots,n-1\} = P^d_{\Delta'}$
- $\int_\Gamma M_{j,d}(z)\,dz = 1$
- with $M_{j,0}(z) = \begin{cases}(z_{j+1}-z_j)^{-1} & ,z\in\Gamma_j \\ 0 & ,z\in\Gamma\setminus\Gamma_j\end{cases}$

  and the recurrence relations

$$\begin{aligned} M_{j,d}(z) &= \frac{d+1}{d}\left(\frac{z-z_j}{z_{j+d+1}-z_j}M_{j,d-1}(z) + \frac{z_{j+d+1}-z}{z_{j+d+1}-z_j}M_{j+1,d+1}(z)\right) \\ M^{(l)}_{j,d}(z) &= \frac{d+1}{z_{j+d+1}-z_j}\left(M^{(l-1)}_{j,d-1}(z) - M^{(l-1)}_{j+1,d-1}(z)\right),\ l\leq d \end{aligned} \tag{3.2}$$

  one has an effective and stable method for the evaluation of B-splines of any degree.

In order to obtain approximate equation (2.2) of (3.1) one has to determine

$$\mathcal{S} M_{j,d+1}(z) = \frac{1}{\pi i}\int_\Gamma \frac{M_{j,d}(\zeta)}{\zeta - z}\,d\zeta$$

at the points z_k on Γ. As mentioned in the introduction an analytic spline is defined by

$$\mathcal{C} v_\Delta(z) = \frac{1}{2\pi i} \int_\Gamma \frac{v_\Delta(\zeta)}{\zeta - z} d\zeta \ , \ z \notin \Gamma, \ v_\Delta \in P^d_{\Delta'} .$$

Hence a basis of the set of analytic splines are the functions $\mathcal{C} M_{j,d}$, $j = 0,\dots,n-1$. These functions are quiet useful for approximation and interpolation of analytic and harmonic functions. Moreover, the complex variable boundary element method of [9] is based on these functions. Theorem 3 gives recurrence relations for $\mathcal{S} M_{j,d}$ and $\mathcal{C} M_{j,d}$ and their derivatives for any degree $d \geq 1$.

Theorem 3: Relations (3.2) remain valid if $M_{j,d}$ is replaced by $\mathcal{S} M_{j,d}$ or $\mathcal{C} M_{j,d}$.

Proof: From (3.2) we have

$$\int_\Gamma \frac{M_{j,d}(\zeta)\, d\zeta}{\zeta - z} = \frac{d+1}{d} \frac{1}{z_{j+d+1} - z_j} \Big(\int_\Gamma \frac{\zeta - z_j}{\zeta - z} M_{j,d-1}(\zeta)\, d\zeta + \int_\Gamma \frac{z_{j+d+1} - \zeta}{\zeta - z} M_{j+1,d-1}(\zeta)\, d\zeta \Big) =$$

$$= \frac{d+1}{d} \frac{1}{z_{j+d+1} - z_j} \Big((z - z_j) \int_\Gamma \frac{M_{j,d-1}(\zeta)}{\zeta - z} d\zeta + \int_\Gamma M_{j,d-1}(\zeta)\, d\zeta + (z_{j+d+1} - z) \int_\Gamma \frac{M_{j+1,d-1}(\zeta)}{\zeta - z} d\zeta - \int_\Gamma M_{j+1,d-1}(\zeta)\, d\zeta \Big).$$

The result follows from $\int_\Gamma M_{j,d}(\zeta)\, d\zeta = 1$. ∎

Remarks:

1. Relations (3.2) and Theorem 3 remain valid if some of the consecutive partition points coincide. Then (3.2) produce complex B-splines with defect > 1 at these points and their singular or Cauchy type integrals.
2. Relation (3.2) and Theorem 3 hold if the contour Γ is Lipschitz.
3. In contrast to the periodic spline methods the collocation with complex splines for equation (3.1) does not need an explicit parametrization of the curve. One has to choose only a certain number of mesh points on Γ and to use that

$$\mathcal{S} M_{j,0}(z) = M_{j,0}(z) + \frac{1}{\pi i (z_{j+1} - z_j)} \ln \frac{z_{j+1} - z}{z_j - z} .$$

References

[1] Ahlberg, J.H.: Splines in the complex plane. In: I.J. Schoenberg (ed.) Approximation with special emphasis on spline functions. New-York, London 1969, 1-27.

[2] Ahlberg, J.H., Nilson, E.N., Walsh, J.L.: Complex cubic splines. Trans. AMS 129 (1967), 391-413.

[3] Ahlberg, J.H., Nilson, E.N., Walsh, J.L.: Properties of analytic splines (1). Complex polynomial splines. J.Math.Anal.Appl. 27 (1969), 262-278.

[4] Ahlberg,J.H., Nilson,E.N., Walsh,J.L.: Complex polynomial splines on the unit circle. J.Math.Anal.Appl. 33 (1971), 234 - 257.

[5] Arnold,D.N., Wendland,W.L.: On the asymptotic convergence of collocation methods. Math.Comp. 41 (1983), 349 - 381.

[6] Elschner,J.: Singular ordinary differential operators and pseudo-differential equations. Berlin 1985.

[7] Han - Lin Chen: Quasiinterpolant splines on the unit circle. J.Appr.Th. 38 (1983), 4, 312 - 318.

[8] Han - Lin Chen,Hvaring,T.: Approximation of complex harmonic functions by complex harmonic splines. Math.Comp. 42 (1984), 151-164.

[9] Hromadka II,T.V.: The complex variable boundary element method. Berlin,Heidelberg,New-York 1984.

[10] Muskhelishvili,N.I.: Some basic problems of the mathematical theory of elasticity. Groningen 1963.

[11] Schmidt,G.: The convergence of Galerkin and collocation methods with splines for pseudodifferential equations on closed curves. ZAA 3 (1984), 371 - 384.

[12] Schmidt,G.: On complex spline interpolation on the unit circle. In: B.Sendov (ed.): Constructive theory of functions '84. Sofia 1984, 803 - 807.

[13] Schmidt,G.: Splines und die näherungsweise Lösung von Pseudodifferentialgleichungen auf geschlossenen Kurven. Inst.f.Math., Report R-Math-09/86, Berlin 1986.

[14] Schoenberg,I.J.: On polynomial spline functions on the unit circle (1). Their interpolatory properties. In: G.Alexits,S.B.Stečkin (eds.): Proceedings of the conference on constructive theory of functions. Budapest 1972, 403 - 418.

[15] Tzimbalario,J.: Interpolation by complex splines. Trans.AMS 243 (1978), 213 - 222.

[16] Wronicz,Z.: On approximation by complex splines. In: B.Sendov (ed.): Constructive function theory '81. Sofia 1983, 577 - 583.

RESULTS AND PROBLEMS IN SHAPE PRESERVING INTERPOLATION AND APPROXIMATION WITH POLYNOMIAL SPLINES

Jochen W. Schmidt *)

1. Introduction

In the last two decades a lot of papers have been published concerned with topics like convex interpolation or convex smoothing of data sets. Also other constraints have been considered, e.g. monotonicity or positivity conditions. It is not the aim of the present paper to give an exhaustive overview. Mainly essential steps for attacking the problems in question shall be pointed out. Several elaborated results some of which are known while others are new will be described as well as intermediate results which are worth to be completed.

In a first step, if possible at all, the problems under consideration are discretized. They are reformulated by finite dimensional systems of inequalities being in general nonlinear. But there are also conditions sufficient only which are linear. For these discretizations criteria are of interest under which a polynomial is nonnegative on a given interval.

In the next step, by solving the finite dimensional systems existence conditions are obtained as well as algorithms for constructing splines with the desired properties.

In addition, a final step is necessary. If existing at all there are in general an infinite number of spline solutions. Thus the question arises to select one of them. As choice function here the mean curvature is taken. For solving the corresponding optimization problems numerically the special structure of the programs is utilized. In general dualization is of advantage since there exist dual programs which are unconstrained.

2. Nonnegativity of Polynomials on Intervals

For a real polynomial p of degree N conditions will be discussed under which

$$p(u) = \sum_{r=0}^{N} a_r u^r \geq 0 \text{ for all } u \in [0,1] \tag{2.1}$$

holds true. Of course,

$$a_r \geq 0, \quad r = 0(1)N \tag{2.2}$$

*) Department of Mathematics, Technical University of Dresden
Mommsenstr. 13, Dresden 8027, German Democratic Republic

is sufficient for (2.1). By substituting $u = \sigma/(1+\sigma)$ it is immediately seen that (2.1) is equivalent to

$$\sum_{r=0}^{N} \alpha_r \sigma^r \geq 0 \text{ for all } \sigma \geq 0, \text{ where } \alpha_r = \sum_{\lambda=0}^{r} \binom{N-\lambda}{r-\lambda} a_\lambda . \tag{2.3}$$

Obviously the requirement

$$\alpha_r \geq 0, \quad r = 0(1)N \tag{2.4}$$

is a sharper sufficient condition for (2.1) than (2.2). For N=1 the latter inequalities are also necessary. Both (2.2) and (2.4) are linear in $a_0,\dots,a_N$ while for $N \geq 2$ conditions which are sufficient and also necessary are no longer linear. For quadratic and cubic polynomials, respectively, hold

Theorem 1, [21]: For $N = 2$ property (2.1) is valid if and only if

$$\alpha_0 \geq 0, \ \alpha_2 \geq 0, \ \alpha_1 \geq -2\sqrt{\alpha_0\alpha_2} . \tag{2.5}$$

Theorem 2, [21]: For $N = 3$ property (2.1) is valid if and only if

$$(\alpha_0,\alpha_1,\alpha_2,\alpha_3) \in A \cup B \tag{2.6}$$

where

$$A = \{(\alpha_0,\alpha_1,\alpha_2,\alpha_3): \alpha_0 \geq 0, \alpha_1 \geq 0, \alpha_2 \geq 0, \alpha_3 \geq 0\},$$

$$B = \{(\alpha_0,\alpha_1,\alpha_2,\alpha_3): \alpha_0 \geq 0, \alpha_3 \geq 0, 4\alpha_0\alpha_2^3 + 4\alpha_3\alpha_1^3 + 27\alpha_0^2\alpha_3^2 - 18\alpha_0\alpha_1\alpha_2\alpha_3 - \alpha_1^2\alpha_2^2 \geq 0\}.$$

For $N \geq 4$ criteria of the above type seem not to be known; in the case $N = 4$ see [14] for an intermediate result.

Analogously, the bipolynomial p will satisfy the condition

$$p(u,v) = \sum_{\mu=0}^{N} \sum_{r=0}^{N} a_{\mu r} u^\mu v^r \geq 0 \text{ for all } (u,v) \in [0,1]\times[0,1] \tag{2.7}$$

if and only if

$$\sum_{\mu=0}^{N} \sum_{r=0}^{N} \alpha_{\mu r} \sigma^\mu \tau^r \geq 0 \text{ for all } \sigma \geq 0, \ \tau \geq 0 \tag{2.8}$$

where

$$\alpha_{\mu r} = \sum_{\varkappa=0}^{\mu} \sum_{\lambda=0}^{r} \binom{N-\varkappa}{\mu-\varkappa}\binom{N-\lambda}{r-\lambda} a_{\varkappa\lambda} . \tag{2.9}$$

This is proved by substituting $u = \sigma/(1+\sigma)$ and $v = \tau/(1+\tau)$. The condition

$$\alpha_{\mu r} \geq 0, \ \mu = 0(1)N, \ r = 0(1)N \tag{2.10}$$

obviously sufficient for (2.7) is also necessary in the case $N = 1$.

3. Shape Preserving Interpolation with Cubic C^1-Splines

3.1. Let $(x_0,y_0),\dots,(x_n,y_n)$ be given data points whose abscissas may define a partition Δ of an interval I, say $I = [0,1]$,

$$\Delta : x_0 = 0 < x_1 < \dots < x_n = 1 . \tag{3.1}$$

By means of the local variable u, $x = x_{i-1} + h_i u$, $h_i = x_i - x_{i-1}$ a cubic spline s can be defined on Δ by

$$s(x) = y_{i-1}+m_{i-1}h_i u+(3\tau_i-2m_{i-1}-m_i)h_i u^2+(m_{i-1}+m_i-2\tau_i)h_i u^3 \tag{3.2}$$

for $0 \le u \le 1$, $i=1(1)n$ where $\tau_i = (y_i-y_{i-1})/h_i$. Then it follows that $s \in C^1(I)$, and

$$s(x_i) = y_i,\ s'(x_i) = m_i,\ i=0(1)n. \tag{3.3}$$

The derivatives $m_0,\dots,m_n$ here are considered to be parameters which are used in order to adapt the spline to constraints.

Applying (2.4) to s" one gets immediately

Proposition 3, [11]: The cubic C^1-spline s defined by (3.2) is convex on I if and only if

$$2m_{i-1}+m_i \le 3\tau_i \le m_{i-1}+2m_i,\ i=1(1)n. \tag{3.4}$$

When starting from s' condition (2.5) leads to

Proposition 4, [6]: A cubic C^1-spline s is monotone increasing on I if and only if

$$m_{i-1} \ge 0,\ m_i \ge 0,\ m_{i-1}-\sqrt{m_{i-1}m_i} + m_i \le 3\tau_i,\ i=1(1)n. \tag{3.5}$$

According to (2.4)

$$m_{i-1} \ge 0,\ m_i \ge 0,\ m_{i-1}+m_i \le 3\tau_i,\ i=1(1)n \tag{3.6}$$

is sufficient for the monotonicity of s on I.

Condition (2.6) implies straightforwardly

Proposition 5, [21]: Under the assumption $y_i \ge 0$, $i=0(1)n$ a cubic C^1-spline s is nonnegative on I if and only if

$$(m_{i-1},m_i) \in W_i,\ i=1(1)n \tag{3.7}$$

where

$$W_i = A_i \cup B_i,\ A_i = \{(x,y):\ h_i x \ge -3y_{i-1}, h_i y \le 3y_i\},$$

$$B_i = \{(x,y):\ 36y_{i-1}y_i(x^2+xy+y^2-3\tau_i(x+y)+3\tau_i^2)-h_i^2x^2y^2+$$
$$+3(y_i x-y_{i-1}y)(2h_i xy-3y_i x+3y_{i-1}y)+4h_i(y_i x^3-y_{i-1}y^3) \ge 0\}.$$

Sufficient for the nonnegativity of s on I is according to (2.4)

$$y_{i-1} \geq 0,\ y_i \geq 0,\ m_{i-1}h_i \geq -3y_{i-1}, m_ih_i \leq 3y_i,\ i=1(1)n. \qquad (3.8)$$

If one is interested in cubic C^2-splines the equalities

$$h_{i+1}m_{i-1}+2(h_i+h_{i+1})m_i+h_im_{i+1} = 3(h_i\tau_{i+1}+h_{i+1}\tau_i),\ i=1(1)n-1 \qquad (3.9)$$

have to be added to (3.4)-(3.8), respectively; see e.g. [16].

For further examples of discretized constraints it is referred to [7], [18],[21],[22],[24], e.g.

3.2 The existence of convex cubic C^1-spline interpolants is equivalent to the solvability of the finite system (3.4). For solving problems of this type in [20],[3] an abstract algorithm is developed which reduces here to

Algorithm 6,[20]: Set $a_0 = -\infty$, $b_0 = +\infty$ or $a_0 = 3\tau_1-2\tau_2$, $b_0 = \tau_1$, and then

$$a_i = \max\{\tau_i,(3\tau_i-b_{i-1})/2\},\ b_i = 3\tau_i-2a_{i-1} \text{ for } i=1(1)n. \qquad (3.10)$$

Theorem 7,[20]: There are convex spline interpolants (3.2) if and only if

$$a_i \leq b_i,\ i=0(1)n. \qquad (3.11)$$

Moreover, if (3.11) is met then all these splines can be constructed by choosing $m_n \in [a_n,b_n]$ and

$$m_{i-1} \in [c_{i-1},d_{i-1}] \text{ for } i=n(-1)1 \qquad (3.12)$$

where

$$c_{i-1} = \max\{a_{i-1},3\tau_i-2m_i\},\ d_{i-1} = \min\{b_{i-1},(3\tau_i-m_i)/2\}.$$

Criterion (3.11) may fail. On the other hand, if (3.11) holds true there exists in general an infinite number of convex interpolants. The problems of monotone and positive interpolation are always solvable with cubic C^1-splines since $m_0 = \ldots = m_n = 0$ satisfies (3.5) and (3.7), respectively.

3.3 For selecting one of the spline interpolants a choice function is introduced. The geometric curvature suggests to use

$$\sum_{i=1}^{n} w_i \int_{x_{i-1}}^{x_i} s''(x)^2dx \to \text{Min!} \qquad (3.13)$$

with $w_i = 1$ or $w_i = 1/(1+\tau_i^2)^3$. Thus, substituting (3.2), in convex interpolation the specially structured programming problem

$$\sum_{i=1}^{n} F_i(m_{i-1},m_i) \to \text{Min! s.t.} (m_{i-1},m_i) \in W_i, i=1(1)n \qquad (3.14)$$

with

$$F_i(x,y) = \frac{4w_i}{h_i}\{(x-\tau_i)^2+(x-\tau_i)(y-\tau_i)+(y-\tau_i)^2\}, \tag{3.15}$$

$$W_i = \{(x,y):\ 2x+y \le 3\tau_i \le x+2y\}$$

arises. As shown in [4] to (3.14) belongs generally the dual program

$$-\sum_{i=1}^{n} H_i^*(p_{i-1},-p_i) \to \text{Max! with } p_o = p_n = 0 \tag{3.16}$$

where the so-called Fenchel conjugate is defined by

$$H_i^*(\xi,\eta) = \sup\{\xi x+\eta y-F_i(x,y):\ (x,y) \in W_i\}. \tag{3.17}$$

In the case (3.15) one obtains with $\sigma_i = h_i/(12w_i)$

$$\begin{aligned} H_i^*(\xi,\eta) &= \tau_i(\xi+\eta)+\sigma_i(\xi^2-\xi\eta+\eta^2) && \text{for } \xi \le 0,\ \eta \ge 0 \\ &= \tau_i(\xi+\eta)+\sigma_i(\xi/2-\eta)^2 && \text{for } 0 \le \xi \le 2\eta \\ &= \tau_i(\xi+\eta)+\sigma_i(\xi-\eta/2)^2 && \text{for } 2\xi \le \eta \le 0 \\ &= \tau_i(\xi+\eta) && \text{for } \xi \ge 2\eta,\ 2\xi \ge \eta. \end{aligned} \tag{3.18}$$

Hence, the dual program (3.16) is unconstrained and can be solved numerically easier than the primal program (3.14),(3.15). In addition, if a solution $(p_o,p_1,\ldots,p_n)$ of (3.16) is known the solution $(m_o,m_1,\ldots,m_n)$ of (3.14) is computed via the return-formula

$$m_{i-1} = \partial_1 H_i^*(p_{i-1},-p_i),\ m_i = \partial_2 H_i^*(p_{i-1},-p_i),\ i=1(1)n, \tag{3.19}$$

see [2]. Analogous results for the monotone and positive interpolation are given in [4],[21] and [22].

4. Convex Interpolation with Quintic C^2-Splines

A quintic spline s on the grid Δ can be defined with $x = x_{i-1}+h_i u$ by

$$\begin{aligned} s(x) &= y_{i-1}+m_{i-1}h_i u + M_{i-1}h_i^2\frac{u^2}{2} + \\ &+(10\tau_i-6m_{i-1}-4m_i + \frac{-3M_{i-1}+M_i}{2}h_i)h_i u^3 + \\ &+(-15\tau_i+8m_{i-1}+7m_i + \frac{3M_{i-1}-2M_i}{2}h_i)h_i u^4 + \\ &+(6\tau_i-3m_{i-1}-3m_i + \frac{-M_{i-1}+M_i}{2}h_i)h_i u^5,\ 0 \le u \le 1,\ i=1(1)n. \end{aligned} \tag{4.1}$$

It is straightforwardly seen that $s \in C^2(I)$, and

$$s(x_i) = y_i,\ s'(x_i) = m_i,\ s''(x_i) = M_i,\ i=0(1)n. \tag{4.2}$$

The convexity requirement on s is discretized by means of theorem 2. Starting from s" the coefficients of the present polynomials (2.3) are

$$\alpha_{oi} = M_{i-1},\ \alpha_{3i} = M_i\ ,$$

$$\alpha_{1i} = (60\tau_i - 36m_{i-1} - 24m_i - 6M_{i-1}h_i + 3M_ih_i)/h_i\ , \tag{4.3}$$

$$\alpha_{2i} = (-60\tau_i + 24m_{i-1} + 36m_i + 3M_{i-1}h_i - 6M_ih_i)/h_i\ .$$

Hence it follows

Proposition 8: A quintic C^2-spline s is convex on I if and only if with (4.3)

$$(\alpha_{oi},\alpha_{1i},\alpha_{2i},\alpha_{3i}) \in A_i \cup B_i,\ i=1(1)n \tag{4.4}$$

where

$$A_i = \{(\alpha_{oi},\alpha_{1i},\alpha_{2i},\alpha_{3i}):\ \alpha_{oi} \geq 0, \alpha_{1i} \geq 0, \alpha_{2i} \geq 0, \alpha_{3i} \geq 0\},$$

$$B_i = \{(\alpha_{oi},\alpha_{1i},\alpha_{2i},\alpha_{3i}):\ \alpha_{oi} \geq 0, \alpha_{3i} \geq 0,\ 4\alpha_{o1}\alpha_{2i}^3 + 4\alpha_{3i}\alpha_{1i}^3 + 27\alpha_{oi}^2\alpha_{3i}^2 - 18\alpha_{oi}\alpha_{1i}\alpha_{2i}\alpha_{3i} - \alpha_{1i}^2\alpha_{2i}^2 \geq 0\}.$$

A criterion on the solvability of the linear system of inequalities

$$\alpha_{oi} \geq 0, \alpha_{1i} \geq 0, \alpha_{2i} \geq 0, \alpha_{3i} \geq 0,\ i=1(1)n, \tag{4.5}$$

which is sufficient for the convexity of s on I, can be derived by a technique proposed in [15].

If $M_o = \ldots = M_n = 0$ is set then a necessary and sufficient condition for the solvability of the reduced systems (4.4) as well as (4.5), namely for

$$3m_{i-1} + 2m_i \leq 5\tau_i \leq 2m_{i-1} + 3m_i,\ i=1(1)n \tag{4.6}$$

reads

$$a_i \leq b_i,\ i=0(1)n. \tag{4.7}$$

Here the numbers $a_o, b_o, \ldots, a_n, b_n$ are to define by

$$a_i = \max\{\tau_i, (5\tau_i - 2b_{i-1})/3\},\ b_i = (5\tau_i - 3a_{i-1})/2,\ i=1(1)n; \tag{4.8}$$

as initial values can be taken $a_o = -\infty$, $b_o = +\infty$, or also $a_o = (5\tau_1 - 3\tau_2)/2$, $b_o = \tau_1$.

5. S-Convex Surface Interpolation with Biquadratic Splines

Let Δ^2 be a rectangular grid on the square $I^2 = [0,1] \times [0,1]$,

$$\Delta^2: \begin{array}{l} x_0 = 0 < x_1 < \dots < x_n = 1 \\ y_0 = 0 < y_1 < \dots < y_m = 1 , \end{array} \tag{5.1}$$

and let $z_{i,j}$, $i=0(1)n$, $j=0(1)m$ be given real numbers. Set $h_i = x_i - x_{i-1}$, $k_j = y_j - y_{j-1}$, and introduce local variables u and v by $x = x_{i-1} + h_i u$ and $y = y_{j-1} + k_j v$, respectively. Then, with $a(u) = 1-u^2$, $b(u) = u^2$ and $c(u) = u(1-u)$ a biquadratic spline s on Δ^2 can be defined by

$$s(x,y) = \begin{bmatrix} a(u) \\ b(u) \\ c(u) \end{bmatrix}^T \begin{bmatrix} z_{i-1,j-1} & z_{i-1,j} & q_{i-1,j-1}k_j \\ z_{i,j-1} & z_{i,j} & q_{i,j-1}k_j \\ p_{i-1,j-1}h_i & p_{i-1,j}h_i & r_{i-1,j-1}h_ik_j \end{bmatrix} \begin{bmatrix} a(v) \\ b(v) \\ c(v) \end{bmatrix}, \tag{5.2}$$

for $0 \le u \le 1$, $0 \le v \le 1$, $i=1(1)n, j=1(1)m$; see e.g. [23]. Obviously s interpolates the given data set,

$$s(x_i,y_j) = z_{i,j}, \quad i=0(1)n, j=0(1)m. \tag{5.3}$$

Further, $s \in C^1(I^2)$ if and only if

$$\begin{array}{l} p_{i,j}+p_{i-1,j} = 2(z_{i,j}-z_{i-1,j})/h_i, \quad i=1(1)n-1, j=0(1)m \\ q_{i,j}+q_{i,j-1} = 2(z_{i,j}-z_{i,j-1})/k_j, \quad i=0(1)n, j=1(1)m-1 \\ r_{i,j}+r_{i,j-1}+r_{i-1,j}+r_{i-1,j-1} = 4(z_{i,j}-z_{i,j-1}-z_{i-1,j}+z_{i-1,j-1})/(h_ik_j), \\ \qquad\qquad i=1(1)n-1, j=1(1)m-1 \end{array} \tag{5.4}$$

holds true, and then one gets

$$\partial_1 s(x_i,y_i) = p_{i,j}, \partial_2 s(x_i,y_j) = q_{i,j}, \ \partial_1\partial_2 s(x_i,y_j) = r_{i,j}. \tag{5.5}$$

Now, a function f of two variables is called S-convex if

$$\partial_1 f(x,y) \le \partial_1 f(u,v), \partial_2 f(x,y) \le \partial_2 f(u,v) \ \text{for } x \le u, y \le v; \tag{5.6}$$

see [12],[13] where, moreover, this notation is compared with other more popular convexity definitions. From (5.6) it follows that a biquadratic C^1-spline s is S-convex on I^2 if and only if

$$\partial_1\partial_1 s(x,y) \ge 0, \partial_2\partial_2 s(x,y) \ge 0, \ \partial_1\partial_2 s(x,y) \ge 0 \tag{5.7}$$

for $x_{i-1} \le x \le x_i$, $y_{j-1} \le y \le y_j$, $i=1(1)n, j=1(1)m$. The first two conditions are quadratic inequalities in v and u, respectively. They are discretized by proposition 1. The third condition leads to a bilinear inequality which is discretized via (2.10). The result is

Proposition 9: A biquadratic C^1-spline s is S-convex on I^2 if and only if

$$\begin{aligned}
&p_{i,j} \geq p_{i-1,j},\ i=1(1)n,\ j=0(1)m, \\
&q_{i,j} \geq q_{i,j-1},\ i=0(1)n,\ j=1(1)m, \\
&k_j(r_{i,j-1}-r_{i-1,j-1}) \geq -2(p_{i,j-1}-p_{i-1,j-1})-2\sqrt{(p_{i,j-1}-p_{i-1,j-1})(p_{i,j}-p_{i-1,j})}, \\
&h_i(r_{i-1,j}-r_{i-1,j-1}) \geq -2(q_{i-1,j}-q_{i-1,j-1})-2\sqrt{(q_{i-1,j}-q_{i-1,j-1})(q_{i,j}-q_{i,j-1})}, \\
&r_{i-1,j-1} \geq 0,\ p_{i-1,j}-p_{i-1,j-1} \geq k_j r_{i-1,j-1}/2,\ q_{i,j-1}-q_{i-1,j-1} \geq h_i r_{i-1,j-1}/2, \\
&z_{i-1,j-1}-z_{i,j-1}-z_{i-1,j}+z_{i,j} \geq (r_{i-1,j}+r_{i,j-1}+r_{i-1,j-1})h_i k_j/4, \\
&\qquad\qquad i=1(1)n,\ j=1(1)m.
\end{aligned} \tag{5.8}$$

Remark that in view of (5.4) and of the relations

$$r_{i,j-1}+r_{i-1,j-1} = 2(q_{i,j-1}-q_{i-1,j-1})/h_i,\ r_{i-1,j}+r_{i-1,j-1} = 2(p_{i-1,j}-p_{i-1,j-1})/k_j$$

which also result from (5.4), the last three sets of inequalities in (5.8) are superfluous.
If the spline s is given, by means of (5.8) it is easily tested if s is S-convex or not. In the other case that S-convex splines are to be constructed the solvability of the nonlinear system (5.4),(5.8) is to be treated and, if possible at all, a solution $(p_{i,j}, q_{i,j}, r_{i,j})$ is to be computed. Then formula (5.2) for determining s applies.

Since system (5.4),(5.8) is in general not uniquely solvable a choice function is in request. But in two-dimensional problems it seems to be an open question which function should be chosen.

If the biquadratic spline s is required to be convex, monotone, or positive inequality (2.10) yield linear discretizations which are sufficient. This also happens if bicubic splines are used; compare with [14].

6. Convexity of Quadratic Histosplines

A quadratic spline s on the grid Δ reads with $x = x_{i-1}+h_i u$, $h_i = x_i - x_{i-1}$

$$s(x) = u_{i-1}+(u_i-u_{i-1})u+(m_{i-1}h_i-u_i+u_{i-1})u(1-u) \tag{6.1}$$

for $0 \leq u \leq 1$, $i=1(1)n$. Then, $s \in C^1(I)$ if and only if

$$(m_{i-1}+m_i)h_i = 2(u_i-u_{i-1}),\ i=1(1)n-1 \tag{6.2}$$

holds true, and in addition

$$s(x_i) = u_i,\ s'(x_i) = m_i,\ i=0(1)n. \tag{6.3}$$

The requirement on histosplines

$$\int_{x_{i-1}}^{x_i} s(x)dx = Y_i, \quad i=1(1)n \tag{6.4}$$

with given quantities $Y_1,\dots,Y_n$ leads to

$$4u_{i-1}h_i+2u_ih_i+m_{i-1}h_i^2 = 6Y_i, \quad i=1(1)n. \tag{6.5}$$

Further, for the convexity of s on I the condition

$$m_{i-1} \le m_i, \quad i=1(1)n \tag{6.6}$$

is necessary and sufficient. Thus, one gets the

<u>Proposition 10:</u> There exist convex quadratic C^1-histosplines if and only if system (6.2),(6.5),(6.6) is solvable, and any solution $u_o,m_o,\dots,u_n,m_n$ leads to a desired spline via (6.1).

Now, in view of (6.2),(6.5) the numbers u_i,m_i are all expressible by u_o and m_o,

$$u_i = \alpha_iu_o+\beta_im_o+\gamma_i, \quad m_i = \delta_iu_o+\varepsilon_im_o+\varkappa_i. \tag{6.7}$$

As it is seen straightforwardly the coefficients have to satisfy

$$2\alpha_i = -4\alpha_{i-1}-h_i\delta_{i-1}, \quad h_i\delta_i = -2h_i\delta_{i-1}-6\alpha_{i-1};$$

$$2\beta_i = -4\beta_{i-1}-h_i\varepsilon_{i-1}, \quad h_i\varepsilon_i = -2h_i\varepsilon_{i-1}-6\beta_{i-1};$$

$$2h_i\gamma_i = -4h_i\gamma_{i-1}-h_i^2\varkappa_{i-1}+6F_i, \quad h_i^2\varkappa_i = -2h_i^2\varkappa_{i-1}-6h_i\gamma_{i-1}+6F_i.$$

In a next step these formulas are decoupled to

$$\begin{aligned} &h_i\delta_{i-1}+2(h_i+h_{i+1})\delta_i+h_{i+1}\delta_{i+1} = 0, \quad \delta_o = 0, \quad h_1\delta_1 = -6;\\ &h_i\varepsilon_{i-1}+2(h_i+h_{i+1})\varepsilon_i+h_{i+1}\varepsilon_{i+1} = 0, \quad \varepsilon_o = 1, \quad \varepsilon_1 = -1;\\ &h_i\varkappa_{i-1}+2(h_i+h_{i+1})\varkappa_i+h_{i+1}\varkappa_{i+1} = 6\left(\frac{F_{i+1}}{h_{i+1}} - \frac{F_i}{h_i}\right), \quad \varkappa_o = 0, \quad \varkappa_1 = \frac{6F_1}{h_1^2}. \end{aligned} \tag{6.8}$$

Then it can be proved that

$$\delta_i-\delta_{i-1} > 0, \quad \varepsilon_i-\varepsilon_{i-1} > 0 \text{ for i even}$$

$$\delta_i-\delta_{i-1} < 0, \quad \varepsilon_i-\varepsilon_{i-1} < 0 \text{ for i odd}$$

holds true. Thus the condition (6.6), in view of (6.7) being

$$(\delta_i-\delta_{i-1})y_o+(\varepsilon_i-\varepsilon_{i-1})m_o+\varkappa_i-\varkappa_{i-1} \ge 0, \quad i=1(1)n$$

yields the following assertion.

Proposition 11: There are convex quadratic C^1-histosplines if and only if

$$P = \{(y_0,m_0):\ a(y_0) \le m_0 \le b(y_0),\ -\infty < y_0 < +\infty\} \neq \emptyset \tag{6.9}$$

where

$$\begin{aligned} a(x) &= \max\{-[(\delta_i-\delta_{i-1})x+\varkappa_i-\varkappa_{i-1}]/(\varepsilon_i-\varepsilon_{i-1}):\ i \text{ even}\} \\ b(x) &= \min\{-[(\delta_i-\delta_{i-1})x+\varkappa_i-\varkappa_{i-1}]/(\varepsilon_i-\varepsilon_{i-1}):\ i \text{ odd}\}. \end{aligned} \tag{6.10}$$

The set P can be recursively computed by a stable algorithm. Using a technique proposed in [16] this is carried out in paper [8]. There also cubic splines will be included.

7. Convex Smoothing with Cubic C^1-Splines

To a given set of data points $(x_0,y_0),\dots,(x_n,y_n)$ a smooth spline shall be constructed which approximates the data set and which is in addition convex. Here, the spline s is proposed to minimize the functional

$$\ell \sum_{i=1}^{n} w_i \int_{x_{i-1}}^{x_i} s''(x)^2 dx + \sum_{i=1}^{n} \{v_{i-1}(s(x_{i-1})-y_{i-1})^2+v_i(s(x_i)-y_i)^2\}, \tag{7.1}$$

see [1],[23]. The choice of the parameter $\ell > 0$ depends on which of the two sums is considered to be of greater and which of minor importance. The convexity of s is placed into the constraints; $w_i, v_i > 0$ weights.

Now, let s be a cubic C^1-spline on the grid Δ, i.e. with $x = x_{i-1}+h_i u$

$$\begin{aligned} s(x) = {} & u_{i-1}+m_{i-1}h_i u+(-3u_{i-1}+3u_i-2m_{i-1}h_i-m_i h_i)u^2+ \\ & +(2u_{i-1}-2u_i+m_{i-1}h_i+m_i h_i)u^3,\ 0 \le u \le 1,\ i=1(1)n. \end{aligned} \tag{7.2}$$

Then it follows

$$s(x_i) = u_i,\ s'(x_i) = m_i,\ i=0(1)n \tag{7.3}$$

and both the u_i's and m_i's are taken as variables. Using these splines the problem of convex smoothing can be discretized, now leading to the program

$$\begin{aligned} & \sum_{i=1}^{n} F_i(u_{i-1},m_{i-1},u_i,m_i) \to \text{Min!} \\ & \text{s.t. } (u_{i-1},m_{i-1},u_i,m_i) \in W_i,\ i=1(1)n \end{aligned} \tag{7.4}$$

with

$$\begin{aligned} F_i(f,x,g,y) = {} & v_{i-1}(f-y_{i-1})^2+v_i(g-y_i)^2+ \\ & + \frac{4w_i\ell}{h_i}\left\{\left(x-\frac{g-f}{h_i}\right)^2+\left(x-\frac{g-f}{h_i}\right)\left(y-\frac{g-f}{h_i}\right)+\left(y-\frac{g-f}{h_i}\right)^2\right\}, \end{aligned} \tag{7.5}$$

$$W_i = \{(f,x,g,y):\ h_i(2x+y) \le 3g-3f \le h_i(x+2y)\},$$

see [17]. The feasible domain of this program is nonempty; e.g., the constraints are satisfied by $y_i = m_i = 0$, $i=0(1)n$. Further the objective function has a positive definite Hessian. Therefore, problem (7.4), (7.5) is for all data sets uniquely solvable.

For solving (7.4),(7.5) numerically, in paper [17] it is proposed to pass to the unconstrained dual program

$$-\sum_{i=1}^{n} H_i^*(z_{i-1},p_{i-1},-z_i,-p_i) \rightarrow \text{Max! with } z_o = p_o = z_n = p_n = 0. \tag{7.6}$$

The Fenchel conjugates

$$H_i^*(\rho,\xi,\sigma,\eta) = \sup\{\rho f+\xi x+\sigma g+\eta y-F_i(f,x,g,y): (f,x,g,y)\in W_i\} \tag{7.7}$$

here are explicitly given by

$$\begin{aligned} H_i^*(\rho,\xi,\sigma,\eta) &= y_{i-1}\rho+y_i\sigma+(y_i-y_{i-1})(\xi+\eta)/h_i + \\ &+(h_i\rho-(\xi+\eta))^2/(4v_{i-1}h_i^2)+(h_i\sigma+(\xi+\eta))^2/(4v_i h_i^2)+G_i^*(\xi,\eta) \end{aligned} \tag{7.8}$$

where with $\sigma_i = h_i/(12w_i\ell)$

$$\begin{aligned} G_i^*(\xi,\eta) &= \sigma_i(\xi^2-\xi\eta+\eta^2) && \text{for } \xi \le 0,\ \eta \ge 0 \\ &= \sigma_i(\xi/2-\eta)^2 && \text{for } 0 \le \xi \le 2\eta \\ &= \sigma_i(\xi-\eta/2)^2 && \text{for } 2\xi \le \eta \le 0 \\ &= 0 && \text{for } \xi \ge 2\eta,\ 2\xi \ge \eta. \end{aligned} \tag{7.9}$$

In addition, if $(z_o,p_o,z_1,p_1,\dots,z_n,p_n)$ is a solution of (7.6) the solution $(u_o,m_o,\dots,u_n,m_n)$ of (7.4),(7.5) is computed via the return-formula

$$\begin{aligned} u_{i-1} &= \partial_1 H_i^*(z_{i-1},p_{i-1},-z_i,-p_i),\quad m_{i-1} = \partial_2 H_i^*(z_{i-1},p_{i-1},-z_i,-p_i) \\ u_i &= \partial_3 H_i^*(z_{i-1},p_{i-1},-z_i,-p_i),\quad m_i = \partial_4 H_i^*(z_{i-1},p_{i-1},-z_i,-p_i), \\ &\qquad i=1(1)n. \end{aligned} \tag{7.10}$$

Finally, the desired smoothing spline then can be determined by (7.2).

In the same manner smoothing under constraints like

$$\delta(x) \le s^{(k)}(x) \le \varepsilon(x) \text{ for } 0 \le x \le 1 \text{ with } k \in \{0,1,2\}$$

and given bounds δ and ε are considered in [10],[19]. For approaches different from that presented here, see e.g. [9],[5].

References

[1] de Boor,C.: A Practical Guide to Splines. Springer-Verlag 1978.

[2] Burmeister,W., W.Heß and J.W.Schmidt: Convex spline interpolants with minimal curvature. Computing 35, 219-229(1985).

[3] Costantini,P.: On monotone and convex spline interpolation. Math.Comput.46, 203-214(1986).

[4] Dietze,S. and J.W.Schmidt: Determination of shape preserving spline interpolants with minimal curvature via dualization. J.Approx.Theory 52, 43-57(1988).

[5] Elfving,T. and L.-E.Andersson: An algorithm for computing constrained smoothing spline functions. Numer.Math.52,583-595(1988).

[6] Fritsch,F.N. and R.E.Carlson: Monotone piecewise cubic interpolation. SIAM J.Numer.Anal. 17, 238-246(1980).

[7] Gregory,J.A.: Shape preserving spline interpolation. Computer-aided design 18, 53-57(1986).

[8] Heß,W. and J.W.Schmidt: Convex histosplines (in preparation).

[9] Irvine,L.D., S.P.Martin and P.W.Smith: Constrained interpolation and smoothing. Constr.Approx. 2, 129-151(1986).

[10] Lehrack,I.: Konvex-konkaves Ausgleichen mit quadratischen und kubischen Splines. Diplomarbeit TU Dresden 1988.

[11] Neuman,E.: Uniform approximation by some Hermite interpolating splines. J.Comput.Appl.Math. 4, 7-9(1978).

[12] Potra,F.A.: On superadditive rates of convergence. Math.Modell. Numer.Anal. 19, 671-685(1985).

[13] Potra,F.A.: Newton-like methods with monotone convergence for solving nonlinear operator equations. Nonlinear Anal.Theory Meth. Appl. 11, 697-717(1987).

[14] Schmidt,F.: Positive 1D- und 2D-Interpolation mit Splines. Diplomarbeit TU Dresden 1988.

[15] Schmidt,J.W.: Convex interval interpolation with cubic splines. BIT 26, 377-387(1986).

[16] Schmidt,J.W.: On convex cubic C^2-spline interpolation. Intern. Series Numer.Math., vol.81, 213-228(1987).

[17] Schmidt,J.W.: An unconstrained dual program for computing convex C^1-spline approximants. Computing 39, 133-140(1987).

[18] Schmidt,J.W.: On shape preserving spline interpolation: existence theorems and determination of optimal splines. Banach Center publ., vol.22, pp.377-388, Polish Scient.Publ. Warsaw 1988.

[19] Schmidt,J.W.: Convex smoothing by splines and dualization. Confer. Proceed. "Numer.Methods Appl.", Sofia 1988.

[20] Schmidt,J.W. and W.Heß: Schwach verkoppelte Ungleichungssysteme und konvexe Spline-Interpolation. Elem.Math. 39, 85-95(1984).

[21] Schmidt,J.W. and W.Heß: Positivity of cubic polynomials on intervals and positive spline interpolation. BIT 28,340-352(1988).

[22] Schmidt,J.W. and W.Heß: Spline interpolation under two-sided restrictions on the derivatives. Z.Angew.Math.Mech. 69(1989).

[23] Späth,H.: Spline-Algorithmen zur Konstruktion glatter Kurven und Flächen. 4.Aufl., Oldenbourg Verlag 1986.

[24] Wewer,U. Non-negative exponential splines. Computer-aided design 20, 11-16(1988).

EXPONENTIAL SPLINE DIFFERENCE SCHEME FOR SINGULAR PETURBATION PROBLEM[#]

Katarina Surla[*], Dragoslav Herceg[*]

1. Introduction

It is well known that exact solutions of singularly perturbed boundary value problems have exponential properties in boundary layers. The cubic spline collocation method and the corresponding difference scheme have an inherent formal cell Reynold's number limitation. In order to avoid these difficulties the solution of the problem

$$\begin{gathered} Ly \equiv -\varepsilon y'' + p(x)y = f(x),\ x \in I = [0,1], \\ y(0) = y(1) = 0,\ p(x) \geq \gamma > 0, \end{gathered} \tag{1}$$

is sought in [4], [8] in the form of the cubic spline which satisfies the collocation equations

$$-\sigma_j y_j'' + p(x_j)y_j = f(x_j)$$

at the points of the grid

$$\Delta:\ 0 = x_0 < x_1 < \cdots < x_n < x_{n+1} = 1,$$

where σ_j is a fitting factor. In this way exponentially features of the exact solution are transfered to spline coefficients and the corresponding difference scheme has a second order uniform convergence on uniform mesh.

The difference schemes derived by the spline in tension have better properties than the ones, derived via cubic spline. In [6] the scheme which has an optimal error estimate in sense [1] is obtained. The estimate of the form $O(h^{\cdot}\min\{h,\sqrt{\varepsilon}\})$ where h is the mesh step width. A collocation method having boundary layer functions as the base of the spline in tension is given in [9]. In this paper we analyse the corresponding difference scheme. The optimal error estimate is given under weakened conditions on the function p(x). Numerical experiments show good results even in the cases when we have no theoretical proof.

---

*) Institute of Mathematics, Dr Ilije Djuričića 4, 21000 Novi Sad, Yugoslavia

#) This reserach was supported partly by NSF and SIZ for Science of SAP Vojvodina through funds made available to the U.S.-Yugoslav Joint Board on Scientific and Technological Cooperation (grant JF 799).

The schemes which achieve a uniform convergence using nonuniform mesh (this mesh must be appropriately chosen), which is very fine in the boundary layers and coarser elsewhere, are considered in
[2], [3], [11]. Using these schemes one cannot obtain the solution at the points given in advance.

Although the spline difference shemes have the same order of the convergence on uniform and on non-uniform meshes, the exponentially fitted spline difference schemes considered in [4], [7] lose this properties.

The uniform convergence of our scheme is obtained under usual conditions on the mesh:

$$\frac{h_i}{h_{i\pm1}} \le M \quad \text{or} \quad \frac{H}{h} \le M,$$

where $h_i = x_{i+1} - x_i$, $h = \min\{h_i : i = 1,2,\dots,n\}$, $H = \max\{h_i : i = 1,2,\dots,n\}$. Here and throughout the paper M denotes any positive constant independent of ε, n and h_j, $j = 0,1,\dots,n$. So, we can obtain numerical solution in the points given in advance. This fact may be important in the numerical solution of nonlocal singularly perturbed boundary value problem.

2. The scheme

The following Theorem 1, [1], [10], gives some properties of the exact solution y(x) of the problem (1), which are incorporated in the construction of our scheme.

Theorem 1. ([1],[10]) *Let* $y \in C^4[0,1]$ *be a solution of the problem* (1). *Then it holds*

$$y(x) = v(x) + w(x) + g(x),$$

$$v(x) = q_0 \exp\left(-x\sqrt{p(0)/\varepsilon}\right), \qquad w(x) = q_1 \exp\left(-(1-x)\sqrt{p(1)/\varepsilon}\right),$$

where q_0 *and* q_1 *are bounded functions of* ε *independent of* x *and*

$$\left|g^{(i)}(x)\right| \le M\left(1 + \varepsilon^{(1-i)/2}\right), \quad i = 0,1,2,3,4.$$

If $p'(0) = p'(1) = 0$ *then*

$$\left|g^{(i)}(x)\right| \le M\left(1 + \varepsilon^{(2-i)/2}\right), \quad i = 0,1,2,3,4.$$

The solution of the problem (1) is sought in the form of the spline in tension e(x),

$$e(x) = e_j(x) = u_{j+1}t + u_j(1-t) + \frac{d_{j+1}}{\rho_j^2}\left(\frac{sh\mu_j t}{sh\mu_j} - t\right) + \frac{d_j}{\rho_j^2}\left(\frac{sh\mu_j(1-t)}{sh\mu_j} - (1-t)\right),\ x \in [x_j, x_{j+1}],$$

where

$$h_j = x_{j+1} - x_j \quad , \quad \mu_j = \rho_j h_j, \qquad t = \frac{x - x_j}{h_j} \ ,$$

and ρ_j is a tension parameter. It is known that

$$e_j(x) \in \text{span}\{1,\ x,\ \exp(\rho_j x),\ \exp(-\rho_j x)\}, \quad x \in [x_j, x_{j+1}].$$

Taking this fact and Theorem 1 into account, we choose

$$\rho_j = \begin{cases} \sqrt{p(0)/\varepsilon} & \text{for } 0 \le j \le [n/2], \\ \sqrt{p(1)/\varepsilon} & \text{for } [n/2] < j \le n. \end{cases}$$

In this way the boundary layer functions v(x) and w(x) become basic functions for $e_j(x)$. The coefficients d_j are determined from the conditions

$$\varepsilon e_j''(x_j) + p(x_j)\, e_j(x_j) = f(x_j), \quad j = 1,2,\dots,n,$$
$$e_0 = 0, \quad e_n(1) = 0.$$

Since $e(x) \in C^2[0,1]$ we have

$$e_j'(x_j) = e_{j-1}'(x_j),\ j = 1,2,\dots,n.$$

Putting $d_j = \big(p(x_j)u_j - f(x_j)\big)/\varepsilon$ in the above equalities we obtain the following difference scheme

$$Ru_j = Qf_j\ , \quad j = 1,2,\dots,n$$
$$u_0 = 0, \quad u_{n+1} = 0, \tag{2}$$

where

$$Ru_j = r_j^- u_{j-1} + r_j^c u_j + r_j^+ u_{j+1}, \quad Qf_j = q_j^- f_{j-1} + q_j^c f_j + q_j^+ u_{j+1},$$

$$r_j^- = \frac{-1}{h_{j-1}} - \left(\frac{\rho_j}{sh\mu_{j-1}} - \frac{1}{h_{j-1}}\right)\frac{p_{j-1}}{\varepsilon\rho_j^2}, \quad r_j^+ = \frac{-1}{h_j} - \left(\frac{\rho_j}{sh\mu_j} - \frac{1}{h_j}\right)\frac{p_{j+1}}{\varepsilon\rho_j^2}$$

$$r_j^c = \frac{1}{h_j} + \frac{1}{h_{j-1}} + \left(\rho_j \operatorname{cth}\mu_j - \frac{1}{h_j} + \rho_j \operatorname{cth}\mu_{j-1} - \frac{1}{h_{j-1}}\right)\frac{p_j}{\varepsilon\rho_j^2}$$

$$q_j^- = \frac{-1}{\varepsilon\rho_j^2}\left(\frac{\rho_j}{\operatorname{sh}\mu_{j-1}} - \frac{1}{h_{j-1}}\right), \quad q_j^+ = \frac{-1}{\varepsilon\rho_j^2}\left(\frac{\rho_j}{\operatorname{sh}\mu_j} - \frac{1}{h_j}\right),$$

$$q_j^c = \frac{1}{\varepsilon\rho_j^2}\left(\rho_j \operatorname{cth}\mu_j - \frac{1}{h_j} + \rho_j \operatorname{cth}\mu_{j-1} - \frac{1}{h_{j-1}}\right)$$

$$f_j = f(x_j), \quad p_j = p(x_j).$$

When it is clear from the context, the subscripts in $r_j^-, r_j^c, r_j^+, q_j^-, q_j^c, q_j^+$ will be omitted. The local truncation error $\tau_j(\varphi)$ of the scheme (2) for an arbitrary sufficiently smooth function $\varphi(x)$ is defined by

$$\tau_j(\varphi) = R\varphi_j - Q(L\varphi)_j.$$

The scheme (2) can be written down in the matrix form

$$Au = F,$$

where A is the matrix of the system (2), u and F are corresponding vectors.

Because of $R(y_j - u_j) = \tau_j(y)$ we have

$$\max_{1\le j\le n} |y_j - u_j| \le \|A^{-1}\|_\infty \max_{1\le j\le n} |\tau_j(y)|.$$

In order to estimate values $|y_j - u_j|$, we shall estimate the truncation error and the norm of the matrix A^{-1}.

3. The norm of the matrix

Since $r_j^c > 0$ we will show that $\Delta_{ij} > 0$, for $1 \le i \le 4$, $1 \le j \le n$, where

$$\Delta_{1j} = r^c - r^- - r^+, \quad \Delta_{2j} = r^c - r^+ + r^-, \quad \Delta_{3j} = r^c + r^+ - r^-, \quad \Delta_{4j} = r^c + r^+ + r^-,$$

and then we shall have that

$$\|A^{-1}\|_\infty \le \frac{1}{\min\limits_{i,j} \Delta_{ij}} .$$

Let $0 \le x \le 0.5$ and let $p(0) = 1$, which does not diminich generality, $|p'(x)| \le L$, $x \in [0,1]$. There exist ξ_1 and ξ_2 such that $x_{j-1} \le \xi_1 \le x_j \le \xi_2 \le x_{j+1}$ and

$$\Delta_{1j} = \frac{2}{h_j} + \frac{2}{h_{j-1}} + p_i \Big[\frac{1}{\sqrt{\varepsilon}} \operatorname{cth} \mu_i - \frac{2}{h_j} - \frac{2}{h_{j-1}} + \frac{1}{\sqrt{\varepsilon}} \operatorname{cth} \mu_{i-1} + \frac{1}{\sqrt{\varepsilon} \operatorname{sh} \mu_{j-1}} + \frac{1}{\sqrt{\varepsilon} \operatorname{sh} \mu_j} \Big]$$

$$+ \; p'(\xi_2)\, h_j \Big(\frac{1}{\sqrt{\varepsilon} \operatorname{sh} \mu_j} - \frac{1}{h_j} \Big) - p'(\xi_1)\, h_{j-1} \Big(\frac{1}{\sqrt{\varepsilon} \operatorname{sh} \mu_{j-1}} - \frac{1}{h_{j-1}} \Big) .$$

Since $\frac{z}{\operatorname{sh} z}\,(1 + \operatorname{ch} z) \ge 2$, the expression in the first parentheses is nonnegative. Thus, $\Delta_{1j} > 0$ if $a^- > 0$ and $a^+ \ge 0$ or $a^- \ge 0$ and $a^+ > 0$ where

$$a^- = \frac{2}{h_j} + p'(\xi_2)\, h_j \Big(\frac{1}{\sqrt{\varepsilon} \operatorname{sh} \mu_j} - \frac{1}{h_j} \Big) \quad \text{and} \quad a^+ = \frac{2}{h_{j-1}} + p'(\xi_1) h_{j-1} \Big(\frac{1}{\sqrt{\varepsilon} \operatorname{sh} \mu_{j-1}} - \frac{1}{h_{j-1}} \Big).$$

This can be achived for suficiently small h_j. For example: $\Delta_{1j} > 0$ if $\max(h_j, h_{j-1}) \le \frac{2}{L}$. In the restt of the paper we shall suppose that $\frac{H}{h} \le M$.

Let $h_j^2 \le \varepsilon$, $h_{j-1} \le \varepsilon$. Then, after some Taylor expansions, we have $\Delta_{1j} \ge \frac{M}{h_j}$.

For $h_j \ge \varepsilon$, $h_{j-1} \ge \varepsilon$ because of $\operatorname{cth} \rho_j \ge M$, $M_1 e^{\rho_j} \le \operatorname{sh} \rho_j \le M_2 e^{\rho_j}$, $M_3 \le \rho_j < \infty$, we have $\Delta_{1j} \ge \frac{M}{\sqrt{\varepsilon}}$. Analysing the cases $h_j^2 \ge \varepsilon$, $h_{j-1}^2 \ge \varepsilon$ and $h_j^2 \ge \varepsilon, h_{j-1}^2 \ge \varepsilon$ we obtain

$$\Delta_{1j} \ge \begin{cases} \frac{M}{\sqrt{\varepsilon}} & \text{for } h_j, h_{j-1} \le \varepsilon \\ \frac{M}{h_j} & \text{otherwise} \end{cases} \tag{3}$$

Similarly we obtain that $\Delta_{ij} > 0$ for $i = 2, 3, 4$ and that the estimate (3) holds for Δ_{ij} , $i = 2,3,4$.

The same holds for $0.5 < x_j < 1$. The difference is only in definition of ρ_j. Thus,

$$\|A^{-1}\|_\infty \le \begin{cases} M\sqrt{\varepsilon} & \text{for } h_j \le \sqrt{\varepsilon},\ j = 0(1)n, \\ M\varepsilon/h & \text{otherwise.} \end{cases} \tag{4}$$

4. Truncation error

According to Theorem 1 the solution of the problem (1) has the form

$$y(x) = v(x) + q_1(x)\ ,\ 0 \le x \le 0.5,$$
$$y(x) = w(x) + q_2(x)\ ,\ 0.5 \le x \le 1\ ,$$

where for i = 1,2 and j = 0,1,2,3,4

$$|q_i^{(j)}(x)| \le \begin{cases} M\,(1 + \varepsilon^{(2-j)/2}), & \text{when } p'(0) = p'(1) = 0, \\ M\,(1 + \varepsilon^{(1-j)/2}), & \text{otherwise.} \end{cases}$$

Let $p'(0) = p'(1) = 0$. Let $\max(h_j^2, h_{j-1}^2) \le \varepsilon$ and $0 \le x_{j+1} \le 0.5$. Since $\tau_j(y) = \tau_j(v) + \tau_j(q_1)$ and $\tau_j(v) = 0$ we have that $\tau_j(y) = \tau_j(q_1)$.

$$\tau_j(q_1) = T_{2j}\, q_{1j}'' + T_{3j}\, q_{1j}''' + \frac{r^- h_{j-1}^4}{4!} y^{IV}(\xi_1) + \frac{h_j^4}{4!} y^{IV}(\xi_2) +$$
$$+ \varepsilon\, q_j^- \frac{h_{j-1}^2}{2} y^{IV}(\xi_3) - p_{j-1} \frac{h_{j-1}^4}{4!} y^{IV}(\xi_4)\, q_j^- +$$
$$+ \varepsilon\, q_j^+ \frac{h_j^2}{2} y^{IV}(\xi_5) - p_{j+1} \frac{h_j^4}{4!} q_j^+ y^{IV}(\xi_6)\ ,$$

where $x_{j-1} \le \xi_i \le x_{j+1}$, $i = 0(1)6$,

$$T_{2j} = \frac{r^+ h_j^2}{2} + \frac{r^- h_{j-1}^2}{2} + \varepsilon\,(q^- + q^+ + q^c) - q^- p_{j-1} \frac{h_{j-1}^2}{2} - q^+ p_{j+1} \frac{h_j^2}{2},$$

$$T_{3j} = \frac{-r^- h_{j+1}^3 + r^+ h_j^3}{3!} - q^- h_{j-1}\, \varepsilon + q^+ h_j \varepsilon + \frac{p_{j-1} q^- h_{j-1}^3}{3!} + p_{j+1} q^+ \frac{h_j^3}{3!}.$$

Using Taylor expansions $\frac{x}{\operatorname{sh} x} = 1 - \frac{x^2}{6} + O(h^4)$ and $x \operatorname{cth} x = 1 + \frac{x^2}{3} + O(h^4)$ we have

$$r^- h_{j-1}^2 + r^+ h_j^2 = h_j + h_{j-1} + O(h_j^3 \rho),$$

$$2\varepsilon(q_j^- + q_j^c + q_j^+) = -h_j - h_{j-1} + O(h_j^3 \rho).$$

Since $\max\{|q_j^-|, |q_j^c|, |q_j^+|\} \le Mh/\varepsilon\ \max\{|r_j^-|, |r_j^+|\} \le Mh^{-1}$ we have $|T_{2j}| \le Mh_j^3\rho$. In a similar way one can obtain that $|T_{3j}| \le Mh_j^4\rho$ and by Theorem 1

$$|\tau_j(g_1)| \le \begin{cases} Mh_j^3/\varepsilon\,, & \text{when } p'(0) = p'(1) = 0, \\ Mh_j^3/\varepsilon^{1.5}, & \text{otherwise.} \end{cases} \tag{5}$$

et $h_j^2 \ge \varepsilon$ and $x_{j+1} \le 0.5$. Now, we use the following form of the truncation error

$$\tau_j(g_1) = \frac{r^- h_{j-1}^2}{2} g''_1(b_1) + \frac{r^+ h_j^2}{2} g''_1(b_2) + q^- p_{j-1} \frac{h_{j-1}^2}{2} g''_1(b_1) - q^+ p_{j+1} \frac{h_j^2}{2} g''_1(b_2)$$

$$- \ \varepsilon\,(q^- g''_1(x_{j-1}) + q^+ g''_1(x_{j+1}) + q^c g''_1(x_j)), \quad x_{j-1} \le b_1 \le x_j \le b_2 \le x_{j+1},$$

and, since $\max\{|q_j^-|, |q_j^+|\} \le Mh_j^{-1}$, $|q^c| \le M\sqrt{\varepsilon}$, $\max\{|r_j^-|, |r_j^+|\} \le Mh_j^{-1}$, we obtain by Theorem 1

$$|\tau_j(g_1)| \le \begin{cases} Mh_j, & \text{when } p'(0) = p'(1) = 0, \\ \dfrac{Mh_j}{\sqrt{\varepsilon}}, & \text{otherwise.} \end{cases} \tag{6}$$

In the case $x_{j-1} \ge 0.5$ we have that $\tau_j(y) = \tau_j(g_2)$ and estimate (6) holds for $\tau_j(g_2)$ also. For the rest part of the interval [0,1] it holds

$$x_{j-1} \le 0.5,\ x_j > 0.5,\ x_{j+1} > 0.5 \quad \text{or} \quad x_{j+1} > 0.5,\ x_{j-1} \le 0.5,\ x_j \le 0.5.$$

Now, by Theorem 1, $|y^{(i)}| \le M$, $i = 0,1,2,3,4$, and it is easy to verify that the estimate (6) holds for $\tau_j(g_1)$ and $\tau_j(g_2)$.

Let $h_j^2 \le \varepsilon$ and $h_{j-1}^2 \ge \varepsilon$. Since $h_j = Mh_{j-1}$ we have $h_j \le \sqrt{\varepsilon} \le Mh_j$ and

$$|\tau_j(y)| \le M_1 h_{j-1} + M_2 h_j^3/\varepsilon \ \le \begin{cases} Mh_j\,, & \text{when } p'(0) = p'(1) = 0, \\ \dfrac{Mh_j}{\sqrt{\varepsilon}}, & \text{otherwise.} \end{cases}$$

The same situation is when $h_j^2 \ge \varepsilon$ and $h_{j-1}^2 \le \varepsilon$. Thus, if $h_j \lesseqgtr \varepsilon$, $j = 0,1,2,\ldots,n$, $\tau_j(y)$ satisfies the estimate (5). In the opposite case it satisfies the estimate (6).

5. Optimal and uniform convergence

Theorem 2. *Let* $y(x) \in C^4[0,1]$ *and* $H \le Mh$. *Let* $[u_0,u_1,\ldots,u_{n+1}]^T$ *be a solution of the system (2). Then*

$$\max_{1\le j\le n} |y(x_j) - u_j| \le \begin{cases} MH\min\{\sqrt{\varepsilon}, H\}, & \textit{when } p'(0) = p'(1) = 0, \\ \dfrac{MH^2}{H+\varepsilon}, & \textit{otherwise.} \end{cases}$$

P r o o f. Since

$$\max_{1\le j\le n} |y_j - u_j| \le \|A^{-1}\|_\infty \max_{1\le j\le n} |\tau_j(y)|$$

we obtain from (3), (5) and (6)

$$\max_{1\le j\le n} |y(x_j) - u_j| \le \begin{cases} MH^2, & \text{when } p'(0) = p'(1) = 0, \\ \dfrac{MH^2}{\sqrt{\varepsilon}}, & \text{otherwise,} \end{cases}$$

if $h_j^2 \le \varepsilon$, $j = 0,1,2,\dots,n$, and

$$\max_{1\le j\le n} |y(x_j) - u_j| \le \begin{cases} MH\sqrt{\varepsilon}, & \text{when } p'(0) = p'(1) = 0, \\ MH, & \text{otherwise,} \end{cases}$$

if $h_j^2 \ge \varepsilon$ $j = 0,1,2,\dots,n$.

In the case that there are i and j such that $h_j^2 < \varepsilon$ and $h_i^2 \ge \varepsilon$ we have $Mh_i^2 \le \varepsilon \le h_i$ and

$$\max_{1\le j\le n} |y(x_j) - u_j| \le \begin{cases} M\varepsilon \le MH^2 & \text{when } p'(0) = p'(1) = 0, \\ M\sqrt{\varepsilon} \le MH & \text{otherwise.} \end{cases}$$

Thus, Theorem 2 is proved. ■

6. Numerical examples

In this section we present results of some numerical experiments using the scheme described in previous sections. Our examples are often used in the literature [1], [2], [10], [11], to compare different codes.

For the first example

$$-\varepsilon^2 y'' + y + \cos^2 \pi x + 2(\varepsilon\pi)^2 \cos 2\pi x = 0, \; x \in [0,1],$$
$$y(0) = y(1) = 0,$$

the exact solution is

$$y(x) = \frac{\exp(-x/\varepsilon) + \exp((x-1)/\varepsilon)}{1 + \exp(-1/\varepsilon)} - \cos^2 \pi x.$$

For the second example

$$-\varepsilon^2 y'' + \big(1 + x(1 - x)\big)y = f(x),\ x \in [0,1],$$
$$y(0) = y(1) = 0$$

the function f(x) is determined in such a way that the exact solution is

$$y(x) = 1 + (x - 1)\exp(-x/\varepsilon) - x\exp((x-1)/\varepsilon).$$

We denote by E_n the maximum of $|y(x_j) - u_j|$, $j = 1,2,\ldots,n$. Here $[u_0,u_1,\ldots,u_{n+1}]^T$ is corresponding numerical solution of the system (2). Also, we define in the usual way the order of convergence Ord for two succesive values of n in Tables 1 and 2with respective errors E_n and E_{2n}:

$$\text{Ord} = \frac{\log E_n - \log E_{2n}}{\log 2}$$

Tables 1 and 2 present the results for numerical solution obtained by our method for first and second example respectively. These results confirm our estimate obtained in Theorem 2.

Table 1.

n \ ε	2^{-2}	2^{-5}	2^{-8}	2^{-10}	2^{-16}	2^{-32}	
4	1.097(-1) -	4.563(-2) -	7.516(-3) -	1.934(-3) -	3.051(-5) -	4.657(-10) -	E_n Ord
8	3.136(-2) 1.807	1.973(-2) 1.210	4.278(-3) 0.813	1.125(-3) 0.782	1.787(-5) 0.772	2.728(-10) 0.772	
16	8.062(-3) 1.960	5.978(-3) 1.723	2.079(-3) 1.041	5.760(-4) 0.966	9.287(-6) 0.944	1.418(-10) 0.944	
32	2.030(-3) 1.990	1.577(-3) 1.992	9.012(-4) 1.206	2.814(-4) 1.033	4.686(-6) 0.987	7.158(-11) 0.986	
64	5.081(-4) 1.998	3.997(-4) 1.980	3.234(-4) 1.478	1.317(-4) 1.096	2.347(-6) 0.998	3.588(-11) 0.997	
128	1.271(-4) 1.999	1.003(-4) 1.995	9.427(-5) 1.779	5.651(-5) 1.220	1.172(-6) 1.002	1.795(-11) 0.999	
256	3.177(-5) 2.000	2.509(-5) 1.999	2.469(-5) 1.933	2.023(-5) 1.482	5.836(-7) 1.005	8.976(-12) 1.000	
512	7.943(-6) 2.000	6.2274(-6) 2.000	6.250(-6) 1.982	5.893(-6) 1.779	2.895(-7) 1.011	4.498(-12) 0.997	
1024	1.985(-6) 2.000	1.570(-6) 2.000	1.567(-6) 1.996	1.543(-6) 1.933	1.425(-7) 1.023	2.252(-12) 0.998	
2048	4.971(-7) 1.998	3.922(-7) 2.000	3.921(-7) 1.999	3.906(-7) 1.982	6.894(-8) 1.047	1.137(-12) 0.986	

Table 1 shows that the convergence order drops from two to one, when ε decreases. This is because of that the order has been calculated with respect to h. Furthermore, it can be seen the approximation error decreases with decreasing ε, which indicates the convergence with respect to the small parameter. This is especially clear from the upper part of the Table 1, which corresponds to greater h values.

The second example does not satisy the condition $p'(0) = p'(1)$. But, irrespective of this, the results in Table 2 indicate the convergence with respect to the small parameter.

Table 2.

n \ ε	2^{-2}	2^{-5}	2^{-8}	2^{-10}	2^{-16}	2^{-32}	
4	9.787(-3)	3.282(-3)	5.153(-5)	3.214(-6)	7.843(-10)	7.105(-15)	E_n
8	2.383(-3)	4.840(-3)	1.104(-4)	6.883(-6)	1.679(-9)	7.105(-15)	
16	5.913(-4)	3.108(-3)	2.314(-4)	1.443(-5)	3.519(-9)	7.334(-10)	
32	1.476(-4)	9.722(-4)	4.652(-4)	2.965(-5)	7.232(-9)	7.280(-10)	
64	3.687(-5)	2.370(-4)	6.474(-4)	6.017(-5)	1.468(-8)	8.213(-10)	
128	9.217(-6)	5.886(-5)	4.009(-4)	1.186(-4)	2.957(-8)	8.739(-10)	
256	2.304(-6)	1.469(-5)	1.235(-4)	1.633(-4)	5.937(-8)	9.020(-10)	
512	5.761(-7)	3.671(-6)	3.001(-5)	1.006(-4)	1.200(-7)	9.165(-10)	
1024	1.48(-7)	9.178(-7)	7.481(-6)	3.093(-5)	2.382(-7)	1.421(-14)	
2048	3.650(-8)	2.295(-7)	1.867(-6)	7.546(-6)	4.766(-7)	6.963(-15)	

References

[1] Doolan, E.P., J.J.H. Miller, W.A. Schilders: Uniform numerical method for problems with initial and boundary layers, Boole Press, Dublin, 1980.

[2] Herceg, D.: Uniform fourth order scheme for a singlar perturbation problem. (to appear).

[3] Herceg, D., R. Vulanović, N. Petrović: Higher order schemes and Richardson extrapolation for singular perturbation problems. Bull. Austral. Math. Soc. 39(1989), 129-139.

[4] Surla, K.: A uniformly convergent spline difference scheme for a self-adjoint singular perturbation problem. Zb. Rad. Prirod.-Mat. Fak. Ser. Mat. 18(1988) (to appear).

[5] Surla, K., M. Stojanović: A solving singularly perturbed boundary value problem by spline in tension. J. Comput. Appl. Math. 23(1988) (to appear).

[6] Surla, K., M. Stojanović: Singularly perturbed spline difference schemes on non equidistant grid. Z. angew. Math. Mech. 68,3(1988), 171-180.

[7] Surla, K.: Numerical solution of singularly perturbed boundary value problems using adaptive spline function approximation. Anal. Theor. Approx. 16(1987), 175-189.

[8] Surla, K., V. Jerković: Spline difference scheme on a non-uniform mesh. Zb. Rad. Prirod.-Mat. Fak. Ser. Mat. 18(1988) (to appear). [9] Surla, K.: A collocation by ,spline in tension. (to appear).

[10] Vulanović, R.: Mesh construction for discretization of singularly perturbed boundary value problems. Ph. D. Thesis, University of Novi Sad, 1986.

[11] Vulanović, R., D. Herceg, N. Petrović: On the extrapolation for singularly perturbed boundary value problem. Computing, 36(1986), 69-79.

FULL AND WEIGHTED UPWIND FINITE ELEMENT METHODS

Lutz Tobiska [1]

1. Introduction

Several upwind finite element and finite difference methods have been developed for solving singularly perturbed boundary value problems in order to guarantee stability of the corresponding discrete problems. Often the convergence of such methods for second order elliptic problems has been studied in H^1-norm or in L^∞-norm [6-7,9-10,12-14] only. In this paper we discuss the properties of a class of weighted upwind finite element methods for solving one- and two-dimensional model problems. In particular we give conditions for the upwind parameters which guarantee the second order L^2-norm convergence for fixed values of the perturbation parameter.

2. A class of upwind methods in the one-dimensional case

Let us consider the singularly perturbed boundary value problem

$$-\varepsilon u''(x) + b(x)u'(x) = f(x) \quad , \quad u(0)=u(1)=0 \quad . \tag{2.1}$$

In a weak formulation (2.1) reads

$$\text{Find } u\in H_0^1(0,1) \text{ such that for all } v\in H_0^1(0,1) \qquad \varepsilon(u',v') + (bu',v) = (f,v) \tag{2.2}$$

where $(.,.)$ denotes the scalar product in $L^2(0,1)$. We use piecewise linear elements on a regular mesh

$$0=x_0<x_1<\ldots<x_N=1 \ , \ h_i=x_{i+1}-x_i \text{ for } i=0,1,\ldots,N-1 \ ,$$

and assume that there are positive constants c_1, c_2 such that $c_1 h \le h_i \le c_2 h$ $(h = \max_i h_i)$. Denoting the corresponding finite element space by V_h we get the usual discrete problem

$$\text{Find } u_h\in V_h \text{ such that for all } v\in V_h \qquad \varepsilon(u_h',v') + (bu_h',v) = (f,v) \ . \tag{2.3}$$

Our upwind method differs from (2.3) by an elementwise application of the trapeziodal formula for numerical integration of the right hand side

$$f_h(v) = \sum_{i=0}^{N-1} \frac{h_i}{2} [f_{i+1}v_{i+1} + f_i v_i]$$

with $\Phi_i = \Phi(x_i)$, $i=0,1,\ldots,N-1$ for a continuous function Φ and a modified discretization of the convective term. Let

---

1) Technische Universität "Otto von Guericke" Magdeburg,
Sektion Mathematik, PSF 124, DDR-3010 Magdeburg

$x_{i+1/2}=(x_{i+1}+x_i)/2$ and $D_o=[x_o,x_{1/2}]$, $D_1=[x_{1/2},x_{3/2}],\dots,D_N=[x_{N-1/2},x_N]$ a secondary decomposition of (0,1). Then, the modified discretization can be obtained in the following way:

$$(bu',v) = \int_0^1 (bu)'vdx - \int_0^1 b'uvdx$$

$$\sim \sum_{i=o}^{N} v_i[b_{i+1/2}u(x_{i+1/2})-b_{i-1/2}u(x_{i-1/2})]-u_iv_i[b_{i+1/2}-b_{i-1/2}] \ .$$

Now we replace $u_h(x_{i+1/2})$ by a linear combination of u_i and u_{i+1}

$$u(x_{i+1/2}) \sim \lambda_i u_i + (1-\lambda_i)u_{i+1}$$

and define the modified bilinear form b_h by

$$b_h(u,v) = \sum_{i=0}^{N-1} (u_{i+1}-u_i)b_{i+1/2}[\lambda_i v_{i+1}+(1-\lambda_i)v_i] \quad , \ u,v\in V_h$$

Then, the weighted upwind method is characterized by

$$\begin{aligned} &\text{Find } u_h\in V_h \text{ such that for all } v\in V_h \\ &\varepsilon(u_h',v') + b_h(u_h,v) = f_h(v) \ . \end{aligned} \tag{2.4}$$

In general the upwind parameter λ_i is chosen as a function of the local Peclet-number $q_{i+1/2}=b_{i+1/2}h_i/(2\varepsilon)$, that means $\lambda_i=\psi(q_{i+1/2})$ for $i=0,1,\dots,N-1$. If we take $\psi(q)=(1+\operatorname{sgn} q)/2$ (2.4) is called a full upwind method.

In case of a uniform mesh ($h=h_i=$const) the weighted upwind method (2.4) corresponds to the finite difference scheme

$$\begin{aligned} -\varepsilon D_+D_-u_i + b_{i+1/2}(\tfrac{1}{2} - \alpha_{i+1/2})D_+u_i + b_{i-1/2}(\tfrac{1}{2} - \alpha_{i-1/2})D_-u_i = f_i \\ i=1,\dots,N-1 \ , \quad u_o = u_N = 0 \end{aligned} \tag{2.5}$$

where $\alpha_{i+1/2} = \lambda_i - \frac{1}{2}$, $i= ,1,\dots,N-1$. Special cases of (2.5) for a constant function b are:

$\alpha(q) = 0$	approximation by central divided differences
$\alpha(q) = \frac{1}{2} \operatorname{sgn} q$	method of Dorr [3]
$\alpha(q) = \frac{1}{2} (\coth q - \frac{1}{q})$	method of Il'in [8]
$\alpha(q) = q/(2 + 2\|q\|)$	method of Samarskij [11]

Replacing the right hand side f_i by

$$(\tfrac{1}{2} + \beta_{i-1/2})f_{i-1/2} + (\tfrac{1}{2} - \beta_{i+1/2})f_{i+1/2}$$

we obtain following schemes:

$\alpha(q) = \beta(q) = \frac{1}{2} \operatorname{sgn} q$	method of Abrahamsson [1]
$\alpha(q) = \frac{1}{2}(\coth q - \frac{1}{q})$, $\beta(q) = 0$	method of Emeljanov [5]

$\alpha(q) = \frac{1}{2}(\coth q - \frac{1}{q})$, $\Phi_{i+1/2}$ replaced by $\frac{1}{2}(\Phi_i + \Phi_{i+1})$

method of El-Mistikawy and Werle [4].

In the following we discuss the properties of the method (2.4).

3. Convergence and stability analysis in the one-dimensional case

Let $\|.\|_k$ and $|.|_k$ denote the norm and seminorm in the space $H^k(\Omega)$, respectively. Uniform estimates can be obtained in the fitted norm

$$\|v\|_\varepsilon = (\varepsilon |v|_1^2 + c_o\|v\|_o^2)^{1/2} \quad , \quad c_o > 0 \quad .$$

First we recall the well-known results for the method (2.3) without upwinding. Assuming $-\frac{1}{2} b'(x) \geq c_o > 0$ we have

$$\|u_h\|_\varepsilon \leq c_o^{-1/2} \|f\|_o \tag{3.1}$$

$$\| u - u_h \|_\varepsilon \leq c\, h\, |u|_2 \tag{3.2}$$

$$\| u - u_h \|_\varepsilon \leq c\, h^2\, \varepsilon^{-3/2}\, |u|_2 \tag{3.3}$$

for the solutions u, u_h of (2.2) and (2.3) , respectively. (3.3) implies the second order L^2-norm convergence for fixed values of the parameter ε. We show by an explicit example that for the full upwind method - (2.4) with $\psi(q) = (1+\mathrm{sgn}\, q)/2$ - u_h converges to u of at most first order for fixed ε.

<u>Example:</u> We consider the boundary value problem

$$-u''(x) + u'(x) = 2x-2 \quad , \quad u(0) = 0 \ , \quad u(1) = 1$$

with the exact solution $u(x) = x^2$. Solving the corresponding discrete problem on a uniform mesh we get

$$u_h(x_i) = x_i^2 + h[\, x_i - \frac{(1+h)^i-1}{(1+h)^N-1}\,] \quad .$$

On V_h we define a norm by

$$|\!|\!| v |\!|\!| = (\, h \sum_{i=o}^{N} (v(x_i))^2 \,)^{1/2}$$

which is eqivalent to the L^2-norm (uniformly with respect to h) . For the interpolation $I_h u$ defined by $I_h u(x_i) = u(x_i)$, $i=0,1,\ldots,N$ we obtain by calculation

$$\lim_{h\to 0} h^{-1} |\!|\!| I_h u - u_h |\!|\!| \neq 0 \quad .$$

Consequently, for some positive constants c_1, c_2 we have

$$c_1 |\!|\!| I_h u - u_h |\!|\!| \leq \|I_h u - u_h\|_o \leq c_2 h^2 |u|_2 + \|u-u_h\|_o \ ,$$

that means u_h converges to u of at most first order in L^2-norm.

Let us study the weighted upwind method (2.4) in more detail.

Lemma 1: We assume that $-\frac{1}{2}b'(x) \geq c_o > 0$ and that there are constants $M>0$ and $\varkappa>0$ independent of ε and h satisfying

$$2\varepsilon q\alpha(q) \geq M h^{\varkappa} \quad . \tag{3.4}$$

Then, the following estimate is valid

$$\varepsilon(v',v') + b_h(v,v) \geq \|v\|_\varepsilon^2 + Mh^{\varkappa}|v|_1^2 \quad \text{for all } v\in V_h . \tag{3.5}$$

Proof: From the definition of b_h it follows

$$b_h(v,v) = \sum_{i=0}^{N-1} b_{i+1/2}\alpha_{i+1/2}(v_{i+1}-v_i)^2 - \frac{1}{2}\sum_{i=0}^{N-1} v_i^2\,(b_{i+1/2}-b_{i-1/2})$$

$$\geq Mh^{\varkappa}|v|_1^2 + c_o \sum_{i=0}^{N-1} v_i^2\,(h_i+h_{i-1})/2$$

$$\geq Mh^{\varkappa}|v|_1^2 + c_o\|v\|_o^2 \quad .$$

Remark: According to Lemma 1 the stabilizing effect of the upwind method is based on the additional term $Mh^{\varkappa}|v|_1^2$ which does not vanish for $\varepsilon\to 0$. In the non-turning point case ($b(x)\geq b_o>0$) we get $\varkappa=1$ for $\alpha(q)=1/2$ and $\varkappa=2$ for the methods of Il'in and Samarskij, respectively.

Now we consider the discretization error caused by the upwinding and by numerical integration.

Lemma 2: Let f and b belong to $W^{2,\infty}(0,1)$ and let α satisfy

$$2\varepsilon q\alpha(q) \leq M h^{\varkappa} \tag{3.6}$$

for some positive constants $\varkappa$ and M. Then, we have the following estimates

$$|\ (f,w) - f_h(w)\ | \leq c\ h^2\ \|f\|_{2,\infty}\ |w|_1 \quad \text{for all } w\in V_h \tag{3.7}$$

$$|\ b_h(v,w) - (bv',w)\ | \leq c\ h^{\min(2,\varkappa)}\ |v|_1\ |w|_1 \quad \text{for all } v,w\in V_h\ . \tag{3.8}$$

Proof: For proving (3.7) we refer to [2]. In order to show (3.8) we estimate $|\ b_h^*(v,w) - (bv',w)\ |$ where b_h^* is defined by

$$b_h^*(v,w) = \frac{1}{2}\sum_{i=0}^{N-1} (u_{i+1}-u_i)(b_iv_i+b_{i+1}v_{i+1}) \quad . \tag{3.9}$$

Numerical integration of (bv',w) by means of trapezoidal rule yields (3.9). Therefore, applying standard arguments of [2] we obtain

$$|\ (bv',w) - b_h^*(v,w)\ )| \leq c\ \|b\|_{2,\infty}|v|_1|w|_1 \quad \text{for all } v,w\in V_h$$

From the definition of b_h and b_h^* we have

$$b_h(v,w)-b_h^*(v,w) = \sum_{i=0}^{N-1} b_{i+1/2}\alpha_{i+1/2}(v_{i+1}-v_i)(w_{i+1}-w_i)$$
$$- \frac{1}{2}\sum_{i=0}^{N-1} (b_{i+1}-b_{i+1/2})(v_{i+1}-v_i)(w_{i+1}-w_i)$$
$$- \frac{1}{2}\sum_{i=0}^{N-1} (b_{i+1}-2b_{i+1/2}+b_i)(v_{i+1}-v_i)w_i$$

Taking into consideration (3.6) and that

$$|b_{i+1}-b_{i+1/2}| \leq \|b\|_{1,\infty} h_i$$

$$|b_{i+1}-2b_{i+1/2}+b_i| \leq \|b\|_{2,\infty} h_i^2$$

we finally conclude

$$|b_h(v,w)-b_h^*(v,w)| \leq M h^{\varkappa}|v|_1|w|_1 + c\|b\|_{2,\infty}h^2|v|_1|w|_1$$

Combining Lemma 1 and Lemma 2 we obtain the following

Theorem 1: Let $\alpha(q)$ satisfy

$$M_1 h^{\varkappa_1} \leq 2\varepsilon q\alpha(q) \leq M_2 h^{\varkappa_2}$$

with positive constants $\varkappa_1$, $\varkappa_2$, M_1, M_2 and let f, b belong to $W^{2,\infty}(0,1)$ with $-\frac{1}{2}b'(x) \geq c_o > 0$. Then the solution u_h of the weighted upwind method (2.4) satisfies the estimates

$$\|u_h\|_{\varepsilon_1} \leq c_o^{-1/2}\|f\|_o \quad , \varepsilon_1 = \varepsilon + M_1 h^{\varkappa_1} \tag{3.10}$$

$$\|u-u_h\|_{\varepsilon_1} \leq c_1 h|u|_2 + c_2 h^{\min(2,\varkappa_2)}\varepsilon_1^{-1/2} \tag{3.11}$$

$$\|u-u_h\|_o \leq c_3 h^2 \varepsilon^{-3/2}|u|_2 + c_4 h^{\min(2,\varkappa_2)}\varepsilon^{-3/2}\varepsilon_1^{-1/2} \tag{3.12}$$

where the constants c_1, c_2, c_3, c_4 are independent of ε and h.

Proof: (3.10) follows from Lemma 1 directly. Modifying the first Lemma of Strang we get

$$\|u-u_h\|_{\varepsilon_1} \leq c \left\{ \inf_{v\in V_h} \left[|u-v|_1 + \sup_{w\in V_h} \frac{|b_h(v,w)-(bv',w)|}{\|w\|_{\varepsilon_1}} \right] + \sup_{w\in V_h} \frac{|(f,w)-f_h(w)|}{\|w\|_{\varepsilon_1}} \right\}.$$

Taking into consideration Lemma 2 and the usual interpolation estimate we have (3.11). By means of duality arguments we obtain first

$$\|u-u_h\|_o \leq \sup_{g\in L^2} \frac{1}{\|g\|_o} \inf_{\varphi\in V_h} \left\{ c\|u-u_h\|_{\varepsilon_1}|\varphi_g-\varphi|_1 + ch^{\min(2,\varkappa_2)}(1+|u_h|_1)\,|\varphi|_1 \right\}$$

where φ_g denotes the solution of the dual problem

$$\varepsilon(v',\varphi_g') + (bv',\varphi_g) = (g,v) \quad \text{for all } v\in H_o^1(0,1)$$

Using (3.11), interpolation estimates and regularity of the solution of the dual problem we conclude (3.12).

Remark: As a consequence of the Theorem for fixed ε we have second order L^2-norm convergence for the method of Il'in or Samarskij provided that $b(x) \geq b_0 > 0$.

Finally, we note a L^∞-stability result for the corresponding finite difference scheme (2.5). Under the assumtions

$b(x) \geq b_o > 0$, $1+2q\alpha(q) > q$ and $q \longrightarrow q \pm (1+2q\alpha(q))$ monotone nondecreasing on $[0,\infty)$

the stability estimate

$$\max_i |u_i| \leq (1+b_o^{-1}) \max_i |f_i|$$

is valid [13]. We note that these conditions are fulfilled for the above given functions α with exception of $\alpha(q) = 0$ (central divided differences).

4. Problems in the two-dimensional case

For simplicity we study the model problem

$$-\varepsilon\Delta u + b.\nabla u = f(x,y) \quad \text{in } \Omega = (0,1)^2 \quad , \quad u|_{\partial\Omega} = 0 \qquad (4.1)$$

with a constant function b. Starting with a uniform mesh of grid points

$$x_i = ih, \; y_i = ih, \quad i = 0,1,\ldots,N \; , \; N = \frac{1}{h}$$

we obtain a triangulation of Friedrichs-Keller-type by connecting the grid points (x_i,y_j) and (x_{i+1},y_{j+1}). The weighted upwind method is based on linear finite elements over each triangle of the mesh. We denote by V_h the finite element space, by v_{ij} the value $v(x_i,y_j)$ for $v \in V_h$ and by D_{ij} the domains

$$[\, x_i - \frac{h}{2}, x_i + \frac{h}{2} \,] \times [\, y_j - \frac{h}{2}, y_j + \frac{h}{2} \,] \qquad i,j = 1,\ldots,N-1$$

which give - together with some modification near the boundary - a second decomposition of Ω. The discretization of the convective term can be derived in the following steps:

(i) lumping and applying Green's theorem

$$(b.\nabla u, v) \sim \sum_{i,j=1}^{N-1} v_{ij} \int_{D_{ij}} \text{div } (bu) \, dD = \sum_{i,j=1}^{N-1} v_{ij} \int_{\partial D_{ij}} b.n \, u \, ds$$

(ii) replacing of u on ∂D_{ij} by a linear combination of neighbouring values of u

$$\begin{aligned} b_h(u,v) = b_1 h & \sum_{i,j=0}^{N-1} (u_{i+1,j} - u_{ij})[\lambda_1 v_{i+1,j} + (1-\lambda_1) v_{ij}] \\ + b_2 h & \sum_{i,j=0}^{N-1} (u_{i,j+1} - u_{ij})[\lambda_2 v_{i,j+1} + (1-\lambda_2) v_{ij}] \end{aligned} \qquad (4.2)$$

Our weighted upwind method is characterized by

Find $u_h \in V_h$ such that for all $v \in V_h$

$$\varepsilon(\nabla u_h, \nabla v) + b_h(u_h, v) = (f,v) \quad . \qquad (4.3)$$

As in the one-dimensional case we have a stability result.

Theorem 2: Let λ_i satisfy $b_i h(\lambda_i - \frac{1}{2}) \geq Mh^{\varkappa}$, i=1,2 with positive constants M and $\varkappa$. Then, for the solution u_h of (4.3) the estimate

$$(\varepsilon + Mh^{\varkappa})|u_h|_1 \leq c\|f\|_o \qquad (4.4)$$

holds where c is independent of ε and h.

Proof: A direct calculation shows $b_h(v,v) \geq Mh^{\varkappa}|v|_1^2$ for all $v \in V_h$, consequently we get (4.4).

Remark: Using the standard Galerkin-method we have instead of (4.4) $\varepsilon|u_h|_1 \leq c\|f\|_o$ only.

Unfortunately, in general λ_i cannot be chosen such that

$$|b_h(u,v)-(b\nabla u,v)| \leq ch^2|u|_1|v|_1 \quad \text{for all } v\in V_h \quad . \qquad (4.5)$$

In order to verify this we calculate the left hand side of (4.5) for $u=\Phi_{kl}$ and $v=\Phi_{k+1,l+1}$ defined by $\Phi_{kl}(x_i,y_j) = \delta_{ik}\delta_{jl}$ and get

$$| b_1+b_2 | \frac{h}{6} \quad .$$

Thus, a better L^2-convergence rate for fixed ε cannot be obtained along the lines of proof of Theorem 1. However, the analysis in the frame of finite difference methods [13] shows ,that for some special cases there are second order error estimates in the discrete L^∞-norm. In this way it is possible to prove

$$\|u-u_h\|_o \leq \|u-I_h u\|_o + c(h^2 \sum_{i,j=1}^{N-1} [(I_h u-u_h)(x_i,y_j)]^2)^{1/2} \leq C_\varepsilon h^2$$

but the more general case is up to now an open problem.

References

[1] Abrahamsson, L.R.: Difference approximations for singular perturbations with a turning point. Dept. of Comput. Sci., Uppsala University, Rep. 58, 1975

[2] Ciarlet, Ph.: The finite element method for elliptic problems. North-Holland. Amsterdam, New York, Oxford 1978

[3] Dorr, F.W.:The numerical solution of singular perturbations of boundary value problems. SIAM J.Numer.Analysis 7(1970),281-313

[4] El-Mistikawy, T.M. and M.J. Werle: Numerical method for boundary layers with blowing - the exponential box scheme. AIAA J. 16(1978),749-751

[5] Емельянов, К.В.: Разностные схемы высокого порядка точности для сингулярно возмущенной краевой задачи. ДУ 17(1981)77,1171-1183

[6] Heinrich, B.: Coercive and inverse-isotone discretization of diffusion-convection problems. Preprint AdW DDR Karl-Weierstrass-Institut, P-MATH-19/88

[7] Ikeda, T.: Maximum principle in finite element models for convection-diffusion phenomena. North-Holland. Math. Studies 76, Lect. Notes in Numerical and Applied Analysis vol. 4, Amsterdam, et. all 1983

[8] Ильин, А.М.: Разностная схема для дифферензиального уравнения с малым параметром при старшей производный. Мат.Заметки 6(1969)2, 237-248

[9] Ohmori, K. and T. Ushijima: A technique of upstream type applied to a linear nonconforming finite element approximation of convective diffusion equations. RAIRO Anal. Numer. 18(1984),309-332

[10] Risch, U.: An upwind FEM for singularly perturbed elliptic problems and local estimates in the L^∞-norm. (appears in Math. Model. Anal. Numer.)

[11] Самарский, А.А.: О монотонных схемах для эллиптических и параболических уравнений в случае несамосопряженного еллиптического оператора. Журн. вычислит. мат. мат.физ. 5(1965)3, 548-551

[12] Schieweck, F. and L. Tobiska: A nonconforming finite element method of upwind type applied to the stationary Navier-Stokes equations. (appears in Math.Model.Anal.Numer.)

[13] Tobiska, L.: Diskretisierungsverfahren zur Lösung singulär gestörter Randwertprobleme. ZAMM 63(1983),115-123

[14] Tobiska, L. and A. Thiele: A weighted upwind finite element method for solving the stationary Navier-Stokes equation. (appears in WZ TU Magdeburg)

FAMILIES OF QUADRATIC AND CUBIC SPLINE DIFFERENCE SCHEMES

ZORICA UZELAC[1], KATARINA SURLA[2]

1. Introduction

We consider singularly perturbed boundary value problem

$$L_\varepsilon y(x) \equiv -\varepsilon y''(x)+g(x)y(x)=f(x), \quad x\in I=[0,1],$$
$$y(0)=A, \quad y(1)=B, \tag{1}$$
$$g(x)\geq g>0, \quad g(x),f(x)\in C^k(I), \quad k\geq 2, \quad A,B,g,\varepsilon\in R, \quad \varepsilon<<1.$$

Under the assumption $g(x)\geq g>0$, L_ε satisfies the maximum principle and has the unique solution which, in general, displays boundary layers, of width $O(\sqrt{\varepsilon})$, at x=0 and x=1 for small ε (see |1|,|2|,|7 |). The following theorem gives some properties of the exact solution and its derivatives.

Theorem 1. Let $g(x)$, $f(x)\in C^k(I)$, $k\geq 2$ in (1). Then

$$y(x)=p_\varepsilon U(x)+s_\varepsilon W(x)+C(x)$$

where $|p_\varepsilon|$, $|s_\varepsilon|\leq M$, $U(x)=\exp(-x\sqrt{g(o)/\varepsilon})$, $W(x)=\exp(-(1-x)\sqrt{g(1)/\varepsilon})$,

$$|C(x)|\leq M, \quad |C^{(i)}(x)|\leq M\varepsilon^{(1-i)/2}, \quad i=1(1)k+2.$$

Let $g'(0)=g'(1)=0$, then

$$|C^{(i)}(x)|\leq N(1+\varepsilon^{1-i/2}), \quad i=0(1)k+2.$$

Proof. See |2| and |7 |.

Writing $i=j(1)k$, $j,k\in N$ we assume that $j\leq k$ and $i\in\{j,j+1,\dots,k-1,k\}$. Through the paper N denotes different constants independent of ε.

Problem (1) was treated numerically in many papers. In |1|,|7|,|10| a special discretisation meshes was used to get uniform convergence. Exponential fitting was used in |2|,|4|,|3|. A cubic spline collocation scheme for (1) when $\varepsilon\equiv 1$ was considered in |8|. Exponentialy fitted implicit spline difference scheme for (1) was devived in |5| by using spline function from $C^2(I)$.

Let us define the "fitted comparison" problem associated with (1) by

$$\tilde{L}_\sigma\tilde{y}(x)\equiv-\tilde{\sigma}(x,\varepsilon)\tilde{y}''(x)+\tilde{g}(x)\tilde{y}(x)=\tilde{f}(x), \quad x\in I$$
$$\tilde{y}(0)=A, \quad \tilde{y}(1)=B, \tag{2}$$

where $\tilde{g}(x)$, $\tilde{f}(x)$ and $\tilde{\sigma}(x,\varepsilon)$ are piecewise polynomial approximations to $g(x)$, $f(x)$ and $\sigma(x,\varepsilon)$ respectively ($\sigma(x,\varepsilon)$ is a fitting factor to be determined). Using collocation method with (2) as the collocation equation and cubic and quadratic splines from $C^1(I)$ as approximation functions, families of spline difference schemes was derived in |6| and |9|. Choosing particular approximations to g(x) and f(x) we can generate the schemes derived, in |5|,|4| and |8|.

The object of this paper is to present sufficient conditions for the second order convergence uniform in ε for these families of schemes.

[1] Fakultet tehničkih nauka, V.Vlahovića 3, 21000 Novi Sad, Yugoslavia
[2] Institut za matematiku, Dr I.Djuričića 3, 21000 Novi Sad, Yugoslavia

2. Derivation of the schemes

Let us define the uniform mesh $\{x_j\}$, $j=0(1)n$ by $x_j=jh$, $j=0(1)n$, n is an positive integer and $h=1/n$ is the mesh length.

Approximations to the exact solution $y(x)$ of (1) we will find in the form of a cubic spline $v(x)$ and a quadratic spline $V(x)$ which, on the each subinterval $I_j=[x_j,x_{j+1}]$, $j=0(1)n-1$, have the following forms

$$v_j(x)=v_j+(x-x_j)v_j^{(1)}+(x-x_j)^2v_j^{(2)}/2+(x-x_j)^3v_j^{(3)}/6,$$
$$V_j(x)=V_j+(x-x_j)V_j^{(1)}+(x-x_j)^2V_j^{(2)}/2.$$

The approximations to $y_j\equiv y(x_j)$ are denoted by v_j and V_j.

The unknown koefficients v_j, $v_j^{(k)}$, $j=0(1)n-1$, $k=1,2,3$ will be found from the conditions:

- $v_j(x)$ satisfies equation (2) at the grid points x_j and x_{j+1}, $j=0(1)n-1$; $v(x)$ satisfies the boundary conditions,
- $v(x)\in C^1(I)$.

The above conditions give the following system of equations:

$$\begin{aligned}
\tilde{L}_\sigma v_j(x)_{x=x_j} &= \tilde{f}_j(x)_{x=x_j}, && j=0(1)n-1,\\
\tilde{L}_\sigma v_j(x)_{x=x_{j+1}} &= \tilde{f}_j(x)_{x=x_{j+1}}, && j=0(1)n-1,\\
v_j(x)_{x=x_{j+1}} &= v_{j+1}(x)_{x=x_{j+1}}, && j=0(1)n-2,\\
v_j'(x)_{x=x_{j+1}} &= v_{j+1}'(x)_{x=x_{j+1}}, && j=0(1)n-2,\\
v_0(0)=A,\quad & v_{n-1}(1)=B.
\end{aligned}\tag{3}$$

The unknown coefficients V_j, $V_j^{(k)}$, $k=1,2$, $j=0(1)n-1$, we will find from the conditions:

- $V(x)$ satisfies equation (2) at the points $x_{j+1/2}\equiv(x_j+x_{j+1})/2$, $j=0(1)n-1$, and the boundary conditions,
- $V(x)\in C^1(I)$,

which gives the system of the same number of equations and unknowns:

$$\begin{aligned}
\tilde{L}_\sigma V_j(x)_{x=x_{j+1/2}} &= \tilde{f}_j(x)_{x=x_{j+1/2}}, && j=0(1)n-1,\\
V_j(x)_{x=x_{j+1}} &= V_{j+1}(x)_{x=x_{j+1}}, && j=0(1)n-2,\\
V_j'(x)_{x=x_{j+1}} &= V_{j+1}'(x)_{x=x_{j+1}}, && j=0(1)n-2,\\
V_0(0)=A,\quad & V_{n-1}(1)=B.
\end{aligned}\tag{4}$$

Let $\tilde{f}_j(x)=\tilde{f}_j=\text{const}$, $\tilde{g}_j(x)=\tilde{g}_j=\text{const}$ and $\tilde{\sigma}_j(x,\varepsilon)=\tilde{\sigma}_j=\text{const}$, $x\in I_j$, $j=0(1)n-1$ be piecewise constant approximations co $f(x)$, $g(x)$ and $\sigma(x,\varepsilon)$. For every fixed j, $j=1(1)n-1$, let us denote by $\tilde{f}_j^+\equiv\tilde{f}_j$, $\tilde{f}_j^-\equiv\tilde{f}_{j-1}$, $\tilde{g}_j^+\equiv\tilde{g}_j$, $\tilde{g}_j^-\equiv\tilde{g}_{j-1}$, $\tilde{\sigma}_j^+=\tilde{\sigma}_j$ and $\tilde{\sigma}_j^-\equiv\tilde{\sigma}_{j-1}$ these approximations over two neighbouring subintervals I_j and I_{j-1}. Then, we can transform (3) (see |9| for details) to get reccurence relations for the coefficients $v_j^{(1)}$, $v_j^{(2)}$ and $v_j^{(3)}$ and the family of implicit difference schemes:

$$\tilde{R}v_j = \tilde{Q}\tilde{f}_j \tag{5}$$

where

$$\tilde{R}v_j \equiv \tilde{r}_j^- v_{j-1} + \tilde{r}_j^c v_j + \tilde{r}_j^+ v_{j+1}, \quad \tilde{Q}\tilde{f}_j \equiv \tilde{q}_j^- \tilde{f}_{j-1} + \tilde{q}_j^+ \tilde{f}_j = \tilde{q}_j^- \tilde{f}_j^- + \tilde{q}_j^+ \tilde{f}_j^+$$

$$\tilde{r}_j^- = 1/\tilde{\gamma}_j^-, \quad \tilde{r}_j^+ = 1/\tilde{\gamma}_j^+, \quad \tilde{q}_j^- = -h/(2\tilde{\sigma}_j^-), \quad \tilde{q}_j^+ = -h/(2\tilde{\sigma}_j^+),$$

$$\tilde{r}_j^c = -(\tilde{r}_j^+ + \tilde{r}_j^-) + \tilde{g}_j^- \tilde{q}_j^- + \tilde{g}_j^+ \tilde{q}_j^+, \quad \tilde{\gamma}_j^\pm = 6h\tilde{\sigma}_j^\pm/(6\tilde{\sigma}_j^\pm - h^2 \tilde{g}_j^\pm).$$

Similarly we can transform (4) (see |6|) and get the following family of quadratic spline difference schemes:

$$\hat{R}v_j = \hat{Q}\tilde{f}_j, \tag{6}$$

where

$$\hat{R}v_j \equiv \hat{r}_j^- v_{j-1} + \hat{r}_j^c v_j + \hat{r}_j^+ v_{j+1}, \quad \hat{Q}\tilde{f}_j \equiv \hat{q}_j^- \tilde{f}_{j-1} + \hat{q}_j^+ \tilde{f}_j = \hat{q}_j^- \tilde{f}_j^- + \hat{q}_j^+ \tilde{f}_j^+$$

$$\hat{r}_j = 1/\hat{\gamma}_j^\pm, \quad \hat{q}_j^\pm = h^2/(2\hat{A}_j^\pm \hat{\gamma}_j^\pm), \quad \hat{r}_j^c = -(\hat{r}_j^- + \hat{r}_j^+) + \tilde{g}_j^- \hat{q}_j^- + \tilde{g}_j^+ \hat{q}_j^+,$$

$$\hat{A}_j^\pm = h^2 \tilde{g}_j^\pm/8 - \tilde{\sigma}_j^\pm, \quad \hat{\gamma}_j^\pm = h(1 - h^2 \tilde{g}_j^\pm/(4\hat{A}_j)).$$

3. Determination of the fitting factor

Let $\hat{R}^c$, $\hat{Q}^c$, $\tilde{R}^c$ and $\tilde{Q}^c$ denote the operators $\hat{R}$, $\hat{Q}$, $\tilde{R}$ and $\tilde{Q}$ for $g(x)=g=\text{const.}$ and $\sigma(x,\varepsilon)=\sigma=\text{const.}$ When it will be clear from the context, the j subscripts in the coefficients of the schemes will be omitted.

Lemma 1. Let $U(x)$ and $W(x)$ be the function defined in Theorem 1. Then

$$\tilde{R}^c U_j \equiv 0, \quad \tilde{R}^c W_j \equiv 0 \quad \text{for} \quad \sigma = \frac{h^2 g}{6}\left(1 + 3/(2\,\mathrm{sh}^2 \tfrac{h}{2}\sqrt{g/\varepsilon})\right),$$

$$\hat{R}^c U_j \equiv \;, \quad \hat{R}^c W_j \equiv 0 \quad \text{for} \quad \sigma = \frac{h^2 g}{8}\left(1 + 2/\mathrm{sh}^2 \tfrac{h}{2}\sqrt{g/\varepsilon}\right).$$

Proof $\tilde{r}_c^- = \tilde{r}_c^+ = r_c$, $\tilde{r}_c^c = -2r_c - hg/\sigma$ for $g(x)=g=\text{const.}$ Thus,

$$\tilde{R}^c U_j \equiv 0 \Leftrightarrow r_c(\exp(h\sqrt{g/\varepsilon}) + \exp(-h\sqrt{g/\varepsilon}) - 2) - hg/\sigma = 0 \Leftrightarrow 4r_c \mathrm{sh}^2 \tfrac{h}{2}\sqrt{g/\varepsilon} - hg/\sigma = 0 \Leftrightarrow$$

$\Leftrightarrow \sigma = \frac{h^2 g}{6}\left(1 + 3/2\,\mathrm{sh}^2 \tfrac{h}{2}\sqrt{g/\varepsilon}\right)$. The others relation we can prove similarly.∎

For $g(x) \neq \text{const}$ we will define the following piecewise constant approximations for fitting factors:

$$\tilde{\sigma}_j^\pm \equiv \frac{h^2 \tilde{g}_j^\pm}{6}\left(1 + 3/2\,\mathrm{sh}^2 \tfrac{h}{2}\sqrt{\tilde{g}_j^\pm/\varepsilon}\right), \tag{7}$$

$$\tilde{\sigma}_j^\pm \equiv \frac{h^2 \tilde{g}_j^\pm}{8}\left(1 + 2/\mathrm{sh}^2 \tfrac{h}{2}\sqrt{\tilde{g}_j^\pm/\varepsilon}\right), \tag{8}$$

for schemes (5) and (6) respectively.

Now, the coefficients of families cubic and quadratic spline difference schemes (5) and (6) respectively are:

$$\tilde{r}_j^\pm = \tilde{r}(\tilde{\rho}_j^\pm), \quad \tilde{q}_j = 2\tilde{\omega}_j^\pm \tilde{r}_j^\pm/\tilde{g}_j^\pm, \quad \tilde{r}_j^c = -(\tilde{r}_j^+ + \tilde{r}_j^-) + \tilde{g}_j^- \tilde{q}_j^- + \tilde{g}_j^+ \tilde{q}_j^+, \tag{9}$$

$$\hat{r}_j^\pm = \hat{r}(\tilde{\rho}_j^\pm), \quad \hat{q}_j^\pm = 2\tilde{\omega}_j^\pm \hat{r}_j^\pm/\tilde{g}_j^\pm, \quad \hat{r}_j^c = -(\hat{r}_j^- + \hat{r}_j^+) + \tilde{g}_j^- \hat{q}_j^- + \tilde{g}_j^+ \hat{q}_j^+, \tag{10}$$

where $\tilde{\rho}_j^{\pm}\equiv\frac{h}{2}\sqrt{\tilde{g}_j^{\pm}/\varepsilon}$, $\tilde{\omega}_j^{\pm}\equiv sh^2\tilde{\rho}_j^{\pm}$, $\tilde{r}(\rho)=3/h(3+2sh^2\rho))$, $\hat{r}(\rho)=1/(h(1+sh^2\rho))$.

The choice of approximations to g(x) and f(x) in (9) and (10) determines the particular scheme. By choosing $\tilde{g}_j^{+}=\tilde{g}_j^{-}=g_j$ and $\tilde{f}_j^{+}=\tilde{f}_j^{-}=f_j$, (9) and (10) becomes precisely the scheme derived in |4|.

Let us now consider the case when $\tilde{g}(x)$ and $\tilde{f}(x)$ in (2) are piecewise linear approximations to g(x) and f(x), $\tilde{\sigma}(x,\varepsilon)$ is unknown fitting factor. Then it is easy to see that a cubic spline v(x) which satisfies (3) belongs to $C^2(I)$. The corresponding difference scheme has the same coefficients as the one derived in |5|. For $\varepsilon=\sigma=1$ the scheme is reduced to the scheme considered in |8|.

4. Consistency error

Let us denote the discretisation of (1) obtained using scheme (9)

$$v_0=A,\quad \tilde{R}v_j=\tilde{Q}\tilde{f}_j,\quad j=1(1)n-1,\quad v_n=B, \tag{11}$$

and the diseretisation obtained using (10)

$$V_0=A,\quad \hat{R}V_j=\hat{Q}\tilde{f}_j,\quad j=1(1)n-1,\quad V_n=B. \tag{12}$$

The consistency errors of discretisations (11) and (12) we shall denote by $\tilde{\tau}_j(y)$ and $\hat{\tau}_j(y)$, respectively. Let us define functions $z(x)\equiv v(x)-y(x)$ and $e(x)=V(x)-y(x)$, then $\tilde{\tau}_j(Y)=\tilde{R}(z_j)$ and $\hat{\tau}_j(y)=\hat{R}(e_j)$.

We shall consider the case when $\tilde{g}_j^{+}(\tilde{g}_j^{-})$ and $\tilde{f}_j^{+}(\tilde{f}_j^{-})$ are piecewise constant approximations to g(x) and f(x) for $x\in I_j(I_{j-1})$ such that

$$\begin{aligned}
&|\tilde{g}_j^{+}-g\ (x_j^{+})|,\quad |\tilde{f}_j^{+}-f\ (x_j^{+})|\le Mh^2,\quad x_j^{+}=x_j+a^{+}h,\\
&|\tilde{g}_j^{-}-g\ (x_j^{-})|,\quad |\tilde{f}_j^{-}-f\ (x_j^{-})|\le Mh^2,\quad x_j^{-}=x_j-a^{-}h,\\
&j=1(1)n-1,\quad 0\le a^{+},\quad a^{-}\le 1,\quad a^{+},a^{-}\in R.
\end{aligned} \tag{13}$$

Lemma 2. Let $y(x)\in C^4(I)$ be the exact solution of (1), then

$$\begin{aligned}
\tilde{\tau}_j(y) = \tilde{R}(z_j) = -\frac{h}{3}(\tilde{F}_j^{1+}/\tilde{\sigma}_j^{+} + \tilde{F}_j^{1-}/(2\tilde{\sigma}_j^{-})+\tilde{F}_j^{2+}/(2\tilde{\sigma}_j^{+}) + \tilde{F}_j^{2-}/\tilde{\sigma}_j^{-}) - \psi_j^{0}(y)/h +\\
+ \psi_{j-1}^{0}(y)/h - \psi_j^{1}(y) + h\psi_j^{2}(y)/6 + h\psi_{j-1}^{2}(y)/3,\quad j=1(1)n-1,
\end{aligned} \tag{14}$$

where

$$\begin{aligned}
\tilde{F}_j^{1+} &= (\tilde{f}_j^{+}-f_j)+(\tilde{\sigma}_j^{+}-\varepsilon)y_j'' + (g_j-\tilde{g}_j^{+})y_j,\\
\tilde{F}_j^{2+} &= (\tilde{f}_j^{+}-f_{j+1}) + (\tilde{\sigma}_j^{+}-\varepsilon)y_{j+1}'' + (g_{j+1}-\tilde{g}_j^{+})y_{j+1},\\
\tilde{F}_j^{1-} &= (\tilde{f}_j^{-}-f_{j-1}) + (\tilde{\sigma}_j^{-}-\varepsilon)y_{j-1}'' + (g_{j-1}-\tilde{g}_j^{-})y_{j-1},\\
\tilde{F}_j^{2-} &= (\tilde{f}_j^{-}-f_j) + (\tilde{\sigma}_j^{-}-\varepsilon)y_j'' + (g_j-\tilde{g}_j^{-})y_j,\\
\psi_j^{k}(y) &= h^{4-k}y^{(4)}(\xi_k)/(4-k)!,\quad x_j<\xi_k<x_{j+1},\quad k=0,1,2,\quad j=0(1)n-1.
\end{aligned}$$

Proof. In order to find $\tilde{R}(z_j)$ and prove the statement of Lemma 2 we consider the corresponding to (3) equations for z(x). The function $z_{j-1}(x)=v_{j-1}(x)-y(x)$, $x\in I_{j-1}$ satisfy the following equations:

$$-\tilde{\sigma}_j^- z''_{j-1} + \tilde{g}_j^- z_{j-1} = \tilde{F}_j^{1-},$$

$$-\tilde{\sigma}_j^-(z''_{j-1}+hz'''_{j-1})+\tilde{g}_j^-(z_{j-1}+hz'_{j-1}+h^2z''_{j-1}/2+h^3z'''_{j-1}/6) = \tilde{E}_j^{2-},$$

$$z_{j-1}+hz'_{j-1}+hz''_{j-1}/2+h^2z'''_{j-1}/6-\psi_{j-1}^0(y) = z_j,$$

$$z'_{j-1}+hz''_{j-1}+h^2z'''_{j-1}/2-\psi_{j-1}^1(y) = z''_j, \quad \tilde{E}_j^{2-} = \tilde{F}_j^{2-}-\tilde{\sigma}_j^-\psi_{j-1}^2(y)+\tilde{g}_j^-\psi_{j-1}^0(y).$$

We can form the equations which satisfies the function $z_j(x)=v_j(x)-y(x)$, $x\in I_j$. Following the procedu,e described in Section 2 for derivation of difference scheme we get (14).∎

Lemma 3. Let $y(x)\in C^3(I)$ be the exact solution of (1), then

$$\hat{\tau}_j(y) = \hat{R}(\hat{e}_j) = -(F_j^++B_j^+2\delta_j^0(y)/h)\hat{q}_j^+-(F_j^-+B_j^-(2\delta_{j-1}^1(y)-2\delta_{j-1}^0(y)/h) - \\ - \tilde{g}_j^-h(\delta_{j-1}^1(y)-2\delta_{j-1}^0(y)/h))\hat{q}_j^-, \quad j=1(1)n-1, \tag{15}$$

where

$$B_j^\pm = \tilde{g}_j^\pm h/8-\tilde{\sigma}_j^\pm/8, \quad \delta_j^k(y)=h^{3-k}y'''(\eta_k)/(3-k)!, \quad x_j<\eta_j<x_{j+1}, \quad k=1,2, \quad j=0(1)n-1,$$

$$\phi_j^k(y)=(h/2)^{3-k}y'''(\xi_k)/(3-k)!, \quad x_j<\xi_j<x_{j+1/2}, \quad k=0(1)2, \quad j=0(1)n-1,$$

$$F_j^+=(-\tilde{f}_j^++f_{j+1/2})-(\tilde{\sigma}_j^+-\varepsilon)y''_{j+1/2}+(\tilde{q}_j^+-g_{j+1/2})y_{j+1/2}+\tilde{\sigma}_j^+\phi_j^2(y)-\tilde{g}_j^+\phi_j^0(y),$$

$$F_j^- = (-\tilde{f}_j^-+f_{k-1/2})-(\tilde{\sigma}_j^--\varepsilon)y''_{k-1/2}+(\tilde{g}_j^--g_{j-1/2})y_{j-1/2}+\tilde{\sigma}_j^-\phi_{j-1}^2(y)-\tilde{g}_j^-\phi_{j-1}^0(y).$$

Proof. See the proof of Lemma 2.∎

Lemma 4. Let $y(x)\in C^4(I)$ be the solution to (1) and let (13) holds where $a^+=a^-=a$. Then

$$|\tilde{\tau}_j(y)|, \quad |\hat{\tau}_j(y)| \le M\frac{h^3}{\varepsilon\sqrt{\varepsilon}} \quad \text{for} \quad h^2\le\varepsilon, \quad j=0(1)n, \tag{16}$$

$$|\tilde{\tau}_j(y)|, \quad |\hat{\tau}_j(y)| \le M\frac{h^3}{\varepsilon} \quad \text{for} \quad h^2\le\varepsilon, \quad g'(0)=g'(1)=0, \quad j=0(1)n. \tag{17}$$

Proof. Using Theorem 1 we get $\tilde{\tau}_j(y)=p_\varepsilon\tilde{\tau}_j(U)+s_\varepsilon\tilde{\tau}_j(W)+\tilde{\tau}_j(C)$. We shall estimate $\tilde{\tau}_j(U)$, $\tilde{\tau}_j(W)$ and $\tilde{\tau}_j(C)$ separately. Let denote (14)

$$\tilde{\tau}_j(y) = \tilde{A}_j(y)+(\psi_{j-1}^0(y)-\psi_j^0(y))/h - \psi_{j-1}^1(y)+h(\psi_j^2(y)-\psi_{j-1}^2(y))/6. \tag{18}$$

When (13) is fullfiled then

$|\tilde{g}_j^+-2g_j+\tilde{g}_j^-|$, $|\tilde{f}_j^+-2f_j+\tilde{f}_j^-| \le Mh^2$. Since $|\tilde{\sigma}_j^+-\tilde{\sigma}_j^-| \le Mh^3$ we get

$$|\tilde{A}_j(C)| \le M\frac{h^3}{\varepsilon}(1+|C''_j|+\max(|C'(\xi_1)|, \quad |C'(\xi_2)|), \quad x_{j-1}<\xi_1<x_1<\xi_2<x_{j+1}.$$

Using Theorem 1, ie. the estimate $|C^{(i)}(x)| \le M\varepsilon^{(1-i)/2}$, $i=1(1)4$, we get

$$|\tilde{\tau}_j(c)| \le |\tilde{A}_j(c)| + M\frac{h^3}{\varepsilon\sqrt{\varepsilon}} \le M\frac{h^3}{\varepsilon\sqrt{\varepsilon}} \quad \text{for} \quad h^2\le\varepsilon, \quad j=1(1)n-1. \tag{19}$$

For $g'(0)=g'(1)=0$, $|c^{(i)}(x)| \le M(1+\varepsilon^{1-i/2})$, $i=0(1)4$. Thus, using (18) we get

$$|\tilde{\tau}_j(c)| \le |\tilde{A}_j(c)| + M\frac{h^3}{\varepsilon} \le M\frac{h^3}{\varepsilon} \quad \text{for} \quad h^2\le\varepsilon, \ g'(0)=g'(1)=0, \ j=1(1)n-1. \tag{20}$$

For $g(x)=g(0)=\text{const}$, $\tilde{\tau}_j^c(U)\equiv0$ and $\tilde{\tau}_j^c(W)\equiv0$ (see Lemma 1). Thus,

$$|\tilde\tau_j(U)|\le|\tilde\tau_j(U)-\tilde\tau_j^c(U)|\le M\frac{h}{\varepsilon}(|\tilde F^{1+}-\tilde F_c^{1+}|+|\tilde F^{1-}-\tilde F_c^{1-}|+|\tilde F^{2+}-\tilde F_c^{2+}|+|\tilde F^{2-}-\tilde F_c^{2-}|)+$$
$$+M\frac{h^2}{\varepsilon^2}(|\tilde\sigma^+-\sigma^c|+|\tilde\sigma^--\sigma^c|)).$$

$\tilde F^{1+}-\tilde F_c^{1+}=(\tilde\sigma^+-\sigma_c)U_j''$, $\tilde F^{1-}-\tilde F_c^{1-}=(\tilde\sigma_j^--\sigma_c)U_{j-1}''$, $\tilde F^{2+}-\tilde F_c^{2+}=(\tilde\sigma^+-\sigma^c)U_{j+1}''$,

$\tilde F^{2-}-\tilde F_c^{2-}=(\tilde\sigma^--\sigma_c)U_j''$. Using Theorem 1 we get

$$|\tilde\tau_j(U)|\le M\frac{h}{\varepsilon^2}(|\tilde\sigma^+-\sigma_c|+|\tilde\sigma^--\sigma_c|).$$

Let us estimate $\tilde\sigma^+-\sigma_c=\frac{h^2}{6}(\tilde g_j^+-g(o))+\varepsilon((\tilde\rho^+)^2/sh^2\tilde\rho^+-\rho_c^2/sh^2\rho_c)$

For $h^2\le\varepsilon$, $(\tilde\rho^+)^2/sh^2\tilde\rho^+=1-(\tilde\rho^+)^2/3+O(\tilde\rho^+)^4)$ (see |2|), thus

$$|\tilde\sigma^+-\sigma_c|\le Mh^2|\tilde g^+-g(o)|.$$

It follows that

$$|\tilde\tau_j(U)|\le Mh^4/\varepsilon^2,\quad \text{for } h^2\le\varepsilon,\quad j=0(1)n. \tag{21}$$

When $g'(0)=0$ then

$|\tilde\sigma_j^+-\sigma_c|\le Mh^2(x_j^+)^2$ and (21) becomes

$$|\tilde\tau_j(U)|\le M\frac{h^5}{\varepsilon^2},\quad \text{for } h^2\le\varepsilon,\quad g'(0)=0,\quad j=0(1)n. \tag{22}$$

Similarly we get

$$|\tilde\tau_j(W)|\le Mh^4/\varepsilon^2,\quad \text{for } h^2\le\varepsilon,\quad j=0(1)n, \tag{23}$$

$$|\tilde\tau_j(W)|\le Mh^5/\varepsilon^2,\quad \text{for } h^2\le\varepsilon,\quad g'(1)=0,\quad j=0(1)n. \tag{24}$$

Estimates (19), (21) and (23) give

$$|\tilde\tau_j(y)|\le Mh^3/(\varepsilon\sqrt{\varepsilon}),\quad \text{for } h^2\le\varepsilon,\quad j=0(1)n.$$

Estimates (20), (22) and (24) give

$$|\tilde\tau_j(y)|\le Mh^3/\varepsilon \quad \text{for } h^2\le\varepsilon,\quad g'(0)=g'(1)=0,\quad j=0(1)n.$$

The estimates for $\hat\tau_j(W)$ we get similarly.

Lemma 5. Let $y(x)\in C^4(I)$ be the solution to (1) and let (13) holds. Then,

$$|\tilde\tau_j(y)|,\quad |\hat\tau_j(y)|\le M,\quad \text{for } \varepsilon\le h^2,\quad j=0(1)n.$$

Proof. $|\tilde\tau_j(y)|\equiv|\tilde Ry_j-\tilde Q\tilde f_j|\le|\tilde Ry_j-\tilde q_j^-f(x_j^-)-\tilde q_j^+f(x_j^+)|+|\tilde q_j^+(\tilde f_j^+-f(x_j^+))|+|\tilde q_j^-(\tilde f_j^--f(x_j^-))|$

For $\varepsilon\le h^2$, $|\tilde q_j^\pm|\le\frac{M}{h}$, $|\hat q_j^\pm|\le\frac{M}{h}$, thus

$$|\tilde\tau_j(y)|\le|\tilde T_j(y)|+Mh \quad\text{and}\quad |\hat\tau_j(y)|\le|\hat T_j(y)|+Mh$$

where we denoted $\tilde T_j(y)\equiv\tilde Ry_j-\tilde q_j^-f(x_j^-)-\tilde q_j^+f(x_j^+)$, $\hat T_j(y)\equiv\hat Ry_j-\hat q_j^-f(x_j^-)-\hat q_j^+f(x_j^+)$.

$|\tilde T_j(C)|$, $|\hat T_j(C)|\le M$ and $|\tilde\tau_j(C)|$, $|\hat\tau_j(C)|\le M$ for $\varepsilon\le h^2$, $j=0(1)n$.

For $g(x)=g(0)=\text{const}$, $\tilde R^cU_j\equiv0$, $\hat R^cU_j\equiv0$ and

$$|\tilde\tau_j(U)|\le|\tilde T_j(U)-\tilde T_j^c(U)|+Mh,\quad |\hat\tau_j(U)|\le|\hat T_j(U)-\hat T_j^c(U)|+Mh.$$

By use of the estimates $Me^{t}\le \mathrm{sh}\,t\le M_1e^{t}$, $M_2\le t<\infty$, and $t^{k}\exp(-t)\le M\exp(-\theta t)$, $\theta\in(0,1)$, $t\in(0,\infty)$ we can estimate

$$|\tilde{r}_j^{+}-r_j^{c}| \le \frac{M}{h}\,\frac{|\tilde{\rho}_j^{+}-\rho_c|}{\min(\omega_j^{+},\omega_c)} \le M\,\frac{h}{\sqrt{\varepsilon}}\,.$$

Similarly we get $|\tilde{r}_j^{-}-r_c|$, $|\tilde{q}_j^{-}-q_c|$, $|\tilde{q}_j^{+}-q_c|\le M\frac{h}{\sqrt{\varepsilon}}$. Using these estimates we get $|\tilde{T}_j(U)-\tilde{T}_j^{c}(U)|\le M$, thus $|\tilde{\tau}_j(U)|\le M$. It is easy to see that $|\hat{\tau}_j(U)|\le M$.

The estimates for $W(x)$ we get similarly to the one for $U(x)$.●

Lemma 6. Let $y(x)\in C^4(I)$ be the exact solution to (1) where $g'(0)=g'(1)=0$. Let (13) holds and let $a^{+}=a^{-}=a$, then

$$|\tilde{\tau}_j(y)|,\quad |\hat{\tau}_j(y)| \le Mh, \quad \text{for } \varepsilon\le h^2,\quad j=0(1)n.$$

Proof. For $g'(0)=0$ and $g(x)=g(0)=\mathrm{const}$ we get $|\tilde{T}_j(U)-\tilde{T}_j^{c}(U)|$, $|\hat{T}_j(U)-\hat{T}_j^{c}(U)| \le Mh$.

For $g'(1)=0$ and $g(x)=g(1)=\mathrm{const}$ we get $|\tilde{T}_j(W)-\tilde{T}_j^{c}(W)|$, $|\hat{T}_j(W)-\hat{T}_j^{c}(W)| \le Mh$.

For the function $C(x)$ it is necessarily to consider the following Taylor expansion for $\tilde{T}_j(y)$:

$$\tilde{T}_j(y) = \tilde{R}y_j-\tilde{q}_j^{-}(L_\varepsilon y)_{x_j^{-}} - \tilde{q}_j^{+}(L_\varepsilon y)_{x_j^{+}} = y_j'\cdot(h(\tilde{r}_j^{+}-\tilde{r}_j^{-})-\tilde{q}_j^{-}g(x_j^{-})(x_j^{-}-x_j)-\tilde{q}_j^{+}g(x_j^{+})(x_j^{+}-x_j) +$$

$$+\frac{h^2}{2}(\tilde{r}_j^{-}y''(\xi_1)+\tilde{r}_j^{+}y''(\xi_2))+\varepsilon(\tilde{q}_j^{-}y''(x_j^{-})+\tilde{q}_j^{+}y''(x_j^{+}))-\tilde{q}^{-}g(x_j^{-})\,\frac{(x_j^{-}-x_j)^2}{2}\,y''(\mu_1) -$$

$$-\tilde{q}_j^{+}g(x_j^{+})(x_j^{+}-x_j)^2y''(\mu_2)/2,\quad x_{j-1}<\xi_1<x_j<\xi_2<x_{j+1},\quad x_j^{-}<\mu_1<x_j<\mu_2<x_j^{+}.$$

Using estimates $|C^{(i)}(x)|\le M(1+\varepsilon^{1-i/2})$, $i=0(1)4$, we get $|\tilde{T}_j(C)|\le Mh$.●

Theorem 2. Let v_j and V_h be approximations to the solution $y(x)\in C^4(I)$ of (1) obtained using (5) and (6) respectively. Let (13) holds and let $a^{+}=a^{-}=a$ then

$$|v_j-y_j|,\quad |V_j-y_j| \le Mh^2/\sqrt{\varepsilon} \quad \text{for } h^2\le\varepsilon, \tag{25}$$

$$|v_j-y_j|,\quad |V_j-y_j| \le Mh \quad \text{for } h^2\ge\varepsilon,\quad j=0(1)n. \tag{26}$$

Proof. Let denote by $\tilde{\tilde{R}}$ and $\hat{\hat{R}}$ the matrix associated with $\tilde{R}$ and $\hat{R}$. Then

$$||\tilde{\tilde{R}}^{-1}|| \le \frac{1}{\min|\tilde{g}_j^{-}h/(2\tilde{\sigma}_j^{-})+\tilde{g}^{+}h/(2\tilde{\sigma}_j^{+})|} \le \begin{cases} M\varepsilon/h, & h^2\le\varepsilon, \quad (27)\\ Mh, & h^2\ge\varepsilon, \quad (28)\end{cases}$$

Using estimates (16) and (27) we get (25). Using Lemma 5 and (28) we get (26).●

Theorem 3. Let the conditions of Theorem 2 be fullfiled and let $g'(0)=g'(1)=0$, then

$$|\tilde{\tau}_j(y)|,\quad |\hat{\tau}_j(y)| \le Mh^2,\quad j=0(1)n.$$

Proof. Using (17), Lemma 6 and estimates for $||\tilde{\tilde{R}}^{-1}||$ and $||\hat{\hat{R}}^{-1}||$ we get the statement of Theorem 3.●

Numerical results

We present the numerical results for the following problem (|2|):

$$-\varepsilon y''(x)+y(x) = \cos^2\pi x+2\varepsilon\pi^2\cos 2\pi x, \quad y(0)=y(1)=0$$

with the exact solution

$$y(x) = (\exp(-(1-x)/\sqrt{\varepsilon})+\exp(-x/\sqrt{\varepsilon})) / (1+\exp(-1/\sqrt{\varepsilon}))-\cos^2\pi x.$$

Table 1. contains the numerical results obtained by using schemes (5) and (6) where $\tilde{g}_j^{\pm}=(g_j+g_{j\pm1})/2$ and $\tilde{f}_j^{\pm}=(f_j+f_{j\pm1})/2$, $j=1(1)n-1$. Table 2. contain the numerical results obtained by using scheme (5) and (6) where $\tilde{g}_j^{\pm}=g_{j\pm1}$ and $\tilde{f}_j^{\pm}=f_{j\pm1}$, $j=1(1)n-1$.

The mesh length $h=1/J$ has succesively halved starting with $J=16$ and ending with $J=1024$. The maximum error et all the mesh points $E_\infty=\max|y_j-\tilde{y}_j|$ is listed under E_∞ ($\tilde{y}_j$ is approximation to y_j). The numerical rate of convergence is determinated as in |2|:

$$\text{rate} \equiv (\ell u Z_{k,\varepsilon}-\ell u Z_{k+1,\varepsilon})/\ell u 2, \text{ where } Z_{k,\varepsilon}=\max_j|\tilde{y}_j^{k/2^k} - \tilde{y}_j^{h/2^{k+1}}|, \quad k=0(1)4,$$

where $\tilde{y}_j^{h/2^k}$ denotes the value of $\tilde{y}_j$ at the mesh point x_j for the mesh length $h/2^k$. The real numbers in tables are given in the form: m(n) wich means $m\cdot10^{-n}$.

$1/\varepsilon$	2	2^2	2^3	2^4	2^5	2^6	2^7	2^8	2^9	2^{10}
K J										
	E∞ rate	E∞ rate	E∞ rate	E∞ rate	E∞ rate	E∞ rate	E∞ rate	E∞ rate	E∞ rate	E∞ rate
0	.22(1)	.21(1)	.18(1)	.16(1)	.14(1)	.13(1)	.13(1)	.13(1)	.13(1)	.13(1)
16	2.00	2.00	2.00	2.00	2.00	2.00	2.01	2.02	2.05	2.10
1	.57(2)	.52(2)	.46(2)	.40(2)	.35(2)	.33(2)	.32(2)	.32(2)	.32(2)	.32(2)
32	2.00	2.00	2.00	2.00	2.00	2.00	2.00	2.00	2.02	2.03
2	.14(2)	.13(2)	.11(2)	.10(2)	.89(3)	.83(3)	.80(3)	.80(3)	.80(3)	.80(3)
64	2.00	2.00	2.00	2.00	2.00	2.00	2.00	2.00	2.00	2.01
3	.36(3)	.33(3)	.29(3)	.25(3)	.22(3)	.20(3)	.20(3)	.20(3)	.20(3)	.20(3)
128	2.00	2.00	2.00	2.00	2.00	2.00	2.00	2.00	2.00	2.00
4	.90(4)	.82(4)	.73(4)	.63(4)	.56(4)	.52(4)	.50(4)	.50(4)	.50(4)	.50(4)
256	2.00	2.00	2.00	2.00	2.00	2.00	2.00	2.00	2.00	2.00
512	.22(4)	.20(4)	.18(4)	.15(4)	.14(4)	.13(4)	.12(4)	.12(4)	.12(4)	.12(4)
1024	.56(5)	.51(5)	.45(5)	.39(5)	.35(5)	.32(5)	.31(5)	.31(5)	.31(5)	.31(5)

T a b l e 1.

$1/\varepsilon$ K	 J	2	2^2	2^3	2^4	2^5	2^6	2^7	2^8	2^9	2^{10}
		E_∞ rate	E_∞ rate	E_∞ rate	E_∞ rate	E_∞ rate	E_∞ rate	E_∞ rate	E_∞ rate	E_∞ rate	E_∞ rate
0		.57(1)	.54(1)	.47(1)	.41(1)	.36(1)	.33(1)	.32(1)	.32(1)	.33(1)	.33(1)
	16	2.00	2.00	2.00	2.00	2.00	2.00	2.00	2.01	2.02	2.04
1		.14(1)	.13(1)	.12(1)	.10(1)	.90(2)	.83(2)	.81(2)	.80(2)	.81(2)	.81(2)
	32	2.00	2.00	2.00	2.00	2.00	2.00	2.00	2.00	2.01	2.01
2		.36(2)	.33(2)	.30(2)	.25(2)	.22(2)	.21(2)	.20(2)	.20(2)	.20(2)	.20(2)
	64	2.00	2.00	2.00	2.00	2.00	2.00	2.00	2.00	2.00	2.00
3		.90(3)	.83(3)	.73(3)	.64(3)	.56(3)	.52(3)	.50(3)	.50(3)	.50(3)	.50(3)
	128	2.00	2.00	2.00	2.00	2.00	2.00	2.00	2.00	2.00	2.00
4		.22(3)	.21(3)	.18(3)	.16(3)	.14(3)	.13(3)	.13(3)	.12(3)	.12(3)	.12(3)
	256	2.00	2.00	2.00	2.00	2.00	2.00	2.00	2.00	2.00	2.00
	512	.56(4)	.52(4)	.46(4)	.40(4)	.35(4)	.32(4)	.32(4)	.31(4)	.31(4)	.31(4)
	1024	.10(4)	.13(4)	.11(4)	.99(5)	.88(5)	.81(5)	.79(5)	.78(5)	.78(5)	.78(5)

T a b l e 2.

References

|1| Bahvalov, N.S.: The optimization of methods of solving boundary value problems with a boundary layer, USSR Computational Math. and Math. Phys., v.9 no. 4, 139-166 (1969)

|2| Doolan, E.P., Miller, J.J.H., Shilders, W.H.A.: Uniform numerical methods for problems with initial and boundary layers, Boole Press, Dublin (1980)

|3| Hegarty, A.F., Miller, J.J.H., O'Riordan, E.: Uniform second order difference schemes for singular perturbation problems, in: J.J.Miller (ed.), BAIL II (1980)

|4| Miller, J.J.H.: On the convergence, uniformly in ε, of difference schemes for a two--point boundary singular perturbation problem, in: Numerical analysis of singular perturbation problems, P.W.Hemker, J.J.H.Miller (ed.), Academic Press (1979)

|5| Surla, K. (to appear in Zb.rad.Prir.Mat.Fak.Univ.u Novom Sadu): A uniformly convergent spline difference scheme for a self-adjoint singular perturbation problem (1988)

|6| Surla, K., Uzelac,Z. (in preparation): The Uniform Convergence of a Class Spline Difference Schemes

|7| Vulanović, R.: Numeričko rešavanje konturnog problema drugog reda sa malim parametrom, Master's Thesis, University of Novi Sad (1983)

|8| Zavyalov, I.S., Kvasov, V.I., Miroshenko: Metody Splyn Funkcii, Nauka, Moskva (1980)

|9| Uzelac, Z., Surla, K (in preparation): A Family of Uniform Convergent Spline Difference Schemes for Self-Adjoint Problems

|10| Uzelac, Z.: On a Fitted Scheme for a Singularly Perturbed Boundary Value Problem, NMAT Conference Proceedings, D.Herceg (ed.) Novi Sad (1985)

Lexikon der Optimierung

Optimierung und optimale Steuerung

Herausgegeben von Alfred Göpfert, Leonhard Bittner, Karl-Heinz Elster, František Nožička, Joachim Piehler und Rainer Tichatschke

1986. VI, 388 Seiten – 14,5 cm × 21,5 cm – Lederin –
DDR 38,– M; Ausland 48,– DM
ISBN 3-05-500031-5
Bestell-Nr. 763 534 5
Bestellwort: Lexikon Optimierung 6919

Ein Schwerpunkt dieses Lexikons liegt neben der genauen Definition von Grundbegriffen auf der Erläuterung numerischer Lösungsverfahren und ihrer Zuordnung zu gewissen Aufgabenklassen der Optimierung bzw. Steuerung. Übersichtsartikel zu einigen Randgebieten und ein viersprachiges Wörterbuch der speziellen Fachausdrücke runden das Werk ab.

Interessenten wenden sich bitte an eine Buchhandlung.

AKADEMIE-VERLAG BERLIN

Band 1

Inverse and Improperly Posed Problems in Differential Equations

Proceedings of the Conference on Mathematical and Numerical Methods, held in Halle/Saale (GDR) from May 29 to June 2, 1979

Edited by Gottfried Anger

296 Seiten, 1979 (vergriffen)

Band 2

Fundamentals of Computation Theory FCT '79

Proceedings of the Conference on Algebraic, Arithmetic and Categorial Methods in Computation Theory, held in Berlin/Wendisch-Rietz (GDR) September 17–21, 1979

Edited by Lothar Budach

576 Seiten, 1979

Band 3

Decidability and Generalized Quantifiers

By Andreas Baudisch, Detlef Seese, Hans-Peter Tuschik, and Martin Weese

XII, 235 Seiten, 1980

Band 4

Zur Elastizitäts- und Thermoelastizitätstheorie

Potentialtheoretische Methoden in Anwendung auf verschiedene Modelle

Von Lothar Jentsch und Johannes Maul

157 Seiten, 1980

Band 5

Systems Analysis and Simulation 1980

Proceedings of the International Symposium held at Berlin (GDR) September 1–5, 1980

Edited by Achim Sydow

496 Seiten, 1980 (vergriffen)

Band 6

Nichtlineare diskrete Optimierung

von Eberhard Girlich und Michail M. Kowaljow

218 Seiten, 1981

Band 7

Completeness Problems in the Structural Theory of Automata

By Jürgen Dassow

148 Seiten, 1981

Band 8

Darstellung von Lösungen linearer elliptischer Differentialgleichungen

Von Günther Wildenhain

92 Seiten, 1981

Band 9

Introduction to the Spectral Theory of Operators in Spaces with an Indefinite Metric

By I. S. Iohvidov, M. G. Krein, and H. Langer

120 Seiten, 1982 (vergriffen)

Band 10

Nonlinear Time-discrete Systems

A General Approach by Nonlinear Superposition

By Michael Gössel

115 Seiten, 1982 (vergriffen)

Band 11

Preprimal Algebras

By Klaus Denecke

162 Seiten, 1982

Band 12

Ergodic Theory and Related Topics

Proceedings of the Conference held in Vitte/Hiddensee (GDR) October 19–23, 1981

Edited by Horst Michel

232 Seiten, 1982

Band 13

Singularly Perturbed Differential Equations

By H. Goering, A. Felgenhauer, G. Lube, H.-G. Roos, and L. Tobiska

176 Seiten, 1983

Band 14

Leben wir in einer Volterra-Welt?

Ein ökologischer Zugang zur angewandten Systemanalyse

Von Manfred Peschel und Werner Mende

233 Seiten, 1983

Band 15

Markov Decision Problems with Countable State Spaces

Optimality Criteria – Algorithms – Clustering

By Hans Michael Dietz and Volker Nollau

174 Seiten, 1983

Band 16

Stochastische Verfahren zur Suche nach dem Minimum einer Funktion

Von Ryszard Zieliński und Peter Neumann

133 Seiten, 1983

Band 17

Invertibility and Asymptotics of Toeplitz Matrices

By Albrecht Böttcher and Bernd Silbermann

200 Seiten, 1983

Band 18

Structural Induction on Partial Algebras

Introduction to Theory and Application of Partial Algebras – Part II

By Horst Reichel

205 pages, 1984

Band 19

Algebraic Methods for Toeplitz-like Matrices and Operators

By Georg Heinig and Karla Rost

212 pages, 1984

Band 20

Frege Conference 1984

Proceedings of the International Conference held at Schwerin (GDR), September 10–14, 1984
Edited by Gerd Wechsung
408 pages, 1984

Band 21

Numerical Analysis of Selected Semilinear Differential Equations

Edited by Christian Großmann, Thomas Riedrich, and Hartmut Schönheinz
203 pages, 1984

Band 22

Singular Ordinary Differential Operators and Pseudodifferential Equations

By Johannes Elschner
200 pages, 1985

Band 23

Lotka-Volterra-Approach to Cooperation and Competition in Dynamic Systems

Proceedings of the 5th Meeting of UNESCO's Working Group on Systems Theory held on the Wartburg, Eisenach (GDR) at March 5–9, 1984
Edited by Werner Ebeling and Manfred Peschel
271 pages, 1985

Band 24

Convergence Structures 1984

Proceedings of the Conference on Convergence held in Bechyně (Czechoslovakia), September 24–28, 1984
Edited by Josef Novák, Werner Gähler, Horst Herrlich, and Jan Mikusiński
254 pages, 1985

Band 25

Parcella '84

Proceedings of the II. International Workshop on Parallel Processing by Cellular Automata and Arrays held in Berlin (GDR), September 25–27, 1984
Edited by Wolfgang Händler, Tamás Legendi, and Gottfried Wolf
239 pages, 1985 (vergriffen)

Band 26

Multiobjective and Stochastic Optimization Based on Parametric Optimization

By Jürgen Guddat, Francisco Guerra Vasquez, Klaus Tammer, and Klaus Wendler
175 pages, 1985

Band 27/28

Systems Analysis and Simulation 1985
I: Theory and Foundations
II: Applications

Proceedings of the International Symposium held in Berlin (GDR), August 26–31, 1985
Edited by Achim Sydow, Manfred Thoma, and Robert Vichnevetsky
per volume 495 pages, 1985

Band 29

Parcella '86

Proceedings of the III. International Workshop on Parallel Processing by Cellular Automata and Arrays held in Berlin (GDR), September 9–11, 1986
Edited by Tamás Legendi, Dennis Parkinson, Roland Vollmar, and Gottfried Wolf
279 pages, 1986

Band 30

Fuzzy Sets Applications, Methodological Approaches, and Results

Proceedings of the International Workshop on Fuzzy Sets Applications cosponsored by IIASA and convened in cooperation with IAEA held on the Wartburg, Eisenach (GDR), at March 3–8, 1985
Edited by Steffen Bocklisch, Sergei Orlovski, Manfred Peschel, and Yasushi Nishiwaki
233 pages, 1986

Band 31

Mathematical Methods of Specification and Synthesis of Software Systems '85

Proceedings of the International Spring School, held in Wendisch-Rietz (GDR), April 22–26, 1985
Edited by Wolfgang Bibel and Klaus P. Jantke
245 pages, 1986 (vergriffen)

Band 32

A Model Theoretic Oriented Approach to Partial Algebras

Introduction to Theory and Application of Partial Algebras – Part I
By Peter Burmeister
319 pages, 1986

Band 33

Finite Difference Method on Irregular Networks

A Generalized Approach to Elliptic Problems of Second Order
By Bernd Heinrich
206 pages, 1986

Band 34

Modern Algorithms for Large Sparse Eigenvalue Problems

By Arnd Meyer
125 pages, 1987

Band 35

Parametric Optimization and Related Topics

Edited by Jürgen Guddat, Hubertus Th. Jongen, Bernd Kummer, and František Nožička
407 pages, 1987

Band 36

Discretization in Differential Equations and Enclosures

Edited by Ernst Adams, Rainer Ansorge, Christian Großmann, and Hans-Görg Roos
250 pages, 1987

Band 37

Factorization of Measurable Matrix Functions

By Georgi Semenovich Litvinchuk and
Ilya Matveyevich Spitkovskiĭ

372 pages, 1987

Band 38

Parallel Algorithms and Architectures

Proceedings of the International Workshop on Parallel Algorithms and Architectures held in Suhl (GDR), May 25–30, 1987

Edited by Andreas Albrecht, Hermann Jung, and Kurt Mehlhorn

205 pages, 1987

Band 39

Parametric Integer Optimization

By Bernd Bank and Reinhard Mandel

138 pages, 1988

Band 40

Computer Analysis of Images and Patterns 1987

Proceedings of the II. International Conference CAIP '87 on Automatic Image Processing, held in Wismar (GDR), September 2–4, 1987

Edited by Leonid P. Jaroslavskiĭ, Azriel Rosenfeld, and Wolfgang Wilhelmi

282 pages, 1987

Band 41

Multivariable Control

A Graph-theoretic Approach

By Kurt J. Reinschke

274 pages, 1988

Band 42

Equilibrium Distributions of Branching Processes

By Alfred Liemant, Klaus Matthes, and Anton Wakolbinger

240 pages, 1988

Band 43

Andreotti-Grauert Theory by Integral Formulas

By Gennadi Markovich Henkin and Jürgen Leiterer

270 pages, 1988

Band 44

Continual Means and Boundary Value Problems in Function Spaces

By Efim Mikhailovich Polishchuk

164 pages, 1988

Band 45

Advances in Mathematical Optimization

Invited papers dedicated to Prof. Dr. Dr. h. c. F. Nožička on occasion of his 70th birthday

Edited by Jürgen Guddat et al.

332 pages, 1988

Band 46/47

Systems Analysis and Simulation 1988

Proceedings of the International Symposium held in Berlin (GDR),

September 12–16, 1988

I: Theory and Foundations
II: Applications

Edited by Achim Sydow, Spyros G. Tzafestas, and Robert Vichnevetsky

426/425 pages, 1988

Band 48

Parcella '88

Proceedings of the IV. International Workshop on Parallel Processing by Cellular Automata and Arrays, held in Berlin (GDR),

October 17–21, 1988

Edited by Gottfried Wolf, Tamás Legendi, and Udo Schendel

379 pages, 1988

Band 49

Algebraic and Logic Programming

Proceedings of an International Workshop held in Gaussig (GDR), November 14–18, 1988
Edited by Jan Grabowski, Pierre Lescanne, and Wolfgang Wechler

278 pages, 1988

Band 50

Asymptotics for Elliptic Mixed Boundary Problems

Pseudo-Differential and Mellin Operators in Spaces with Conormal Singularity

By Stephan Rempel and Bert-Wolfgang Schulze

418 pages, 1989

Band 51

Geobild '89

Proceedings of the 4th Workshop on Geometrical Problems of Image Processing held in Georgenthal (GDR), March 13–17, 1989

Edited by Albrecht Hübler, Werner Nagel, Brian David Ripley, and Günter Werner

198 pages, 1989

The present volume is devoted to spline functions which play an excellent part in Numerical Analysis, as a tool as well as a separate branch.
The volume compiled in preparation of the seminar ISAM–89 collects surveys and more extensive original articles on actual problems in the application of splines in Numerical Analysis.
There are contributions to the following topics: Smoothing by splines, shape preserving interpolation, surface approximation, as well as application of splines in finite-element and collocation methods.
This book is addressed to readers interested in results orientated to applications and also theoretically founded.

Spline-Funktionen, denen der vorliegende Band gewidmet ist, spielen in der Numerischen Mathematik eigenständig oder als Hilfsmittel eine hervorragende Rolle.
Der in Vorbereitung des Seminars ISAM–89 erstellte Band faßt Übersichts- und längere Originalartikel über derzeit aktuelle Probleme in der Anwendung von Splines bei numerischen Aufgabenstellungen zusammen.
Im Vordergrund stehen Beiträge zum Ausgleichen durch Splines, zur Shape-preserving-Interpolation und zur Flächenapproximation sowie zur Verwendung von Splines in Finite-Element- und Kollokationsmethoden.
Die Sammlung wendet sich an Leser, welche an anwendungsorientierten, theoretisch abgesicherten Ergebnissen interessiert sind.

ISBN 3-05-500679-8
ISSN 0138-3019

Bei Fragen zur Produktsicherheit wenden Sie sich bitte an:
If you have any questions regarding product safety,
please contact:

Walter de Gruyter GmbH
Genthiner Straße 13
10785 Berlin
productsafety@degruyterbrill.com